ULTRASONIC IMAGING

DEDICATION

This book is dedicated with great affection and gratitude to my father, Prof. Pál Greguss, Sr., to his 90th birthday, and to my wife Edith. Their faith and loyalty have been a never failing source of inspiration and strength.

ULTRASONIC IMAGING

SEEING BY SOUND

The principles and widespread applications of image formation by sonic, ultrasonic and other mechanical waves

Pál Greguss
M.Sc. (Chemical Engineering),
Ph.D., University of Szeged

Focal Press Limited, London

Focal Press Inc, New York

🕮 British Library Cataloguing in Publication Data

Greguss, Pal
  Ultrasonic imaging.
  1. Acoustic imaging
  I. Title
  621.36′7   TA1550

ISBN 0 240 51039 9 ✓

First edition 1980

Printed in Great Britain by
Thomson Litho Ltd., East Kilbride, Scotland.

CONTENTS

PREFACE

Seeing the invisible was, is and will always be the dream of mankind. Seeing by sound which penetrates through opaque materials is a challenging problem to both engineers and medical people. This book has been prepared in the hope that it may prove to be helpful for those who have the feeling that their scientific, engineering or medical problem could be solved if sound could be used as the information carrier.

The terms 'to see' and 'image' will be defined broadly. If the information pattern from an insonified target can be recorded in such a way that the response to it is in a manner comparable to the response which could be expected if it had been illuminated, we feel the term 'seen' is justified. Therefore, a deliberate attempt has been made to keep the treatment as general as possible, and hence applicable to certain sound image problems to be solved. This is, however, not only a review book of the increasing number of sometimes really very astounding processes which are already here or just around the corner, but it wishes to emphasize that all of these methods are subjected to much the same fundamental principles and limitations, and that psychophysical factors should never be neglected. It attempts to stress always the basic ideas of one or another method, and, therefore, mathematical details are restricted to the minimum, so that those not having the necessary mathematical background may also find the way toward understanding the nature of sound imaging. Based on this understanding the author hopes that some of the readers will wish to explore sound images further, and this is why each chapter has its own independent set of literature references. To please those who are interested in only a limited area of acoustic imaging, the six chapters have been kept fairly self-contained.

In the effort to achieve these goals, the book is organized in the following manner. Chapter 1 presents a brief historical review of the endeavors of seeing by sound, i.e. acoustic imaging. In Chapter 2 background material is given on the relevance of the information theory to sound images, while in Chapter 3 sound waves as information carriers are discussed. Chapter 4 reveals the problems associated with the formation of non sampled and sampled sound images, and in Chapter 5 the different modalities for sound images are discussed. Chapter 6 presents selected material in the application of sound imaging in the different fields of science, engineering and biomedical applications. The question whether the feasibility of seeing sound images by exploiting the possibilities rendered by sensory substitution exists, or not is disucssed at the end of this chapter.

I am indebted to those persons who in some way have helped me to arrive at an understanding of the topic and wish to thank all concerned. I must apologize to those that are not mentioned by name but I am limited by the available space. I entreat their understanding.

Professor Albert Szent-Györgyi first excited my interest in ultrasonics more than 40 years ago when he told me about his pioneering experiment on disintegrating micromolecules with ultrasonic waves, and then he encouraged me as his student at the University of Szeged to work in this field. I owe much to Professors Sandor Szalay and Tamas Tarnoczy, who generously shared with me their experience and knowledge and helped me to build my first ultrasonic generator in 1947.

My interest in acoustic imaging started with the visit of the late Professor Sergey Yakovlevich Sokolov, who encouraged me to start to investigate the problems of sonosensitive materials. Subsequently, I learned much on acoustic imaging from the late Professor Lazar Davidovich Rozenberg, and from Professor Ignacy Malecki, who called my attention to the necessity of always keeping an eye on the non-linear effects of sound propagation. I would also like to thank Professor Raymond W. B. Stephens who provided me with invaluable advice whenever I visited his laboratory.

I am especially grateful to the late Professor Dennis Gabor who was generous in helping me to an understanding of the philosophy of holography which led me to the recording of the first acoustic holograms.

In North America Dr. Floyd Dunn, the late Dr. William Fry, Dr. Adrianus Korpel, Dr. Alex F. Metherell, Dr. Rolf K. Mueller, Dr. Frederick Thurstone, Dr. Gilbert Baum, Professor Dennis N. White were equally willing to help me to obtain a better understanding of the immense field of acoustic imaging, and to guide me from error.

I also wish to thank my collaborators in the laboratories where I worked in the past decade, at the New York Medical College, at the Department of Physics of the Technische Hochschule Darmstadt and the Department of Coherent Optics at the Gesellschaft für Strahlen- und Umweltforschung. Everyone who worked with me has contributed in various ways to this book, but I must especially record my gratitude to Professors Miles A. Galin and Wilhelm Waidelich. Their complete identification with my efforts has been of inestimable value.

My greatest debt, however, I owe to my wife and secretary, Edith, who not only typed without complaint the mountains of typescripts, but also listened patiently to and recognized parts of the book where clarity was in question.

Pal Greguss, Jr.

Memory = sonic ultra → supersonic.
Sound ∿ supersonic
Light = supersonic magic
↓
MACH 2 (y sound inclusive)

1 HISTORICAL INTRODUCTION

Life is possible because systems called organisms can recall and process information reaching them via scattered waves. Objects, for instance, scatter light-borne information which are recorded by photoreceptors, and are processed by a mechanism called 'vision', while 'hearing' is a result of evaluating scattered mechanical waves picked up by mechano-receptors. In the animal world, the perceived information is handled by a network with metabolic and nervous subsystems. There is a main difference, however, between animal life and human life. Animals live without 'knowing' how they live, and they communicate without 'knowing' how they communicate, whereas we not only speculate about how we live and how we communicate but we also try to extend our information gathering capacities beyond our natural capabilities, which in some cases are far behind those of some animals. So, for example, man uses acoustic energy of lower frequencies as a vital means of communication, but this sort of information is never picture-like: the sensing of sonic images is outside the realm of our normal experience. But dolphins and bats were using acoustic energy of higher frequencies, i.e. ultrasonic waves, for getting picture-like information from their environment long before man appeared on Earth.

The suggestion that sound can be used to form images is not really surprising, since the wave propagation characteristics of sound are similar to those of light, only the human information system is unable to decode sound-borne information into a two-dimensional information picture. In the animal world, however, there exists an information processing technique which renders at least a two-dimensional information pattern, conveyed by acoustic waves.

Echolocating animals such as bats[1], dolphins[2], owls and some other birds distinguish between objects of different shapes, recognize whether rods of a grid are horizontal or vertical, and can also distinguish whether the object detected is an obstacle to be avoided or is a desirable food. Measuring the travelling time of their echo can give information, however, only about distance, and cannot answer the problems cited above, since they are related to 'imaging'. These animals therefore have to have an information processing technique similar to vision, but based on sound as the information carrier[3].

The reason for this apparent backwardness of mankind is rather simple. Man uses for information exchange low frequency mechanical waves, i.e. waves which have long wavelength, and so even if man could 'see' these long wavelength created images, as he does light-borne images, these sound images would not only be of low resolution but also be different from that he sees with his eyes. So, for instance, surfaces which have a rough appearance when seen by light may appear quite smooth to these long acoustic wavelengths, a spherical object would appear as a point 'highlight', while the appearance of a cylindrical object would show up as a line 'highlight'.

Nevertheless, since—according to the Chinese proverb—a single picture is worth a thousand words it is not at all surprising that a lot of effort has been devoted by scientists and artists to represent sound-borne information in a two-dimensional form, to create a sonic image, a sonogram. To avoid further misunderstandings, it has to be emphasized that by this no such visual representation of sound as an optical soundtrack of a sound motion picture is meant, since it is not a true picture of sound. This is because it does not give us significant information of the geometry of the sound source and the generated sound field intensity distribution.

The only inherent information processing mechanism in mankind which is able to handle a two-dimensional information pattern and evaluate it as a picture is the visual system, so that the optical replica of the acoustic image has to be presented to the observer for evaluation. This visual image, however, can be regarded as nothing more or less than a link in the communication system which relays the 'invisible' to the human observer as faithfully as possible. When that visualized (two-dimensional) spatial acoustic pattern is interpreted by the eye-brain system, then, and only then, does a sound image have a meaning. This, however, requires an *a priori* knowledge of the interaction of sound with matter, and familiarity with the material to be investigated.

Although the same geometric wave propagation laws can be applied to the image forming process by acoustic waves as in optics, the search for an adequate acoustical-to-optical converter started only in the first decade of this century. The simple reason for this is that until the technical ability of producing high frequency sound waves, i.e. acoustic waves, at least one magnitude shorter than the general size of the object to be 'pictured', there was no hope of succeeding.

Seeing by sound as a technology can probably be said to have had its birth as a consequence of the investigations started by P. Langevin[4] to combat submarine attack during World War I. He realized that the piezoelectric effect discovered some 30 years earlier could be used to generate high intensity ultrasonic vibrations in water, the wavelengths of which were short enough for conveying useful information about objects of interest. The technique he proposed, and which was really put into practice in World War II, was at first sight similar to that used by bats and dolphins: echoranging. The information pattern obtained congruently by this technique was, however, only one-dimensional, so it was unsuited to provide an intensity distribution of the sound field scattered by the object of interest, i.e. its image. This type of seeing by sound gives us only range information.

Sokolov[5] was perhaps the first to solve the problem of recording concurrently sound field intensity distribution, i.e. producing a sonic image, when he demonstrated

12

in 1929 that a liquid surface will be deformed according to the intensity distribution of the ultrasonic field acting on it, and that this deformation can be visualized by reflecting light from the surface, as shown in Fig. 1. The recordings obtained by this

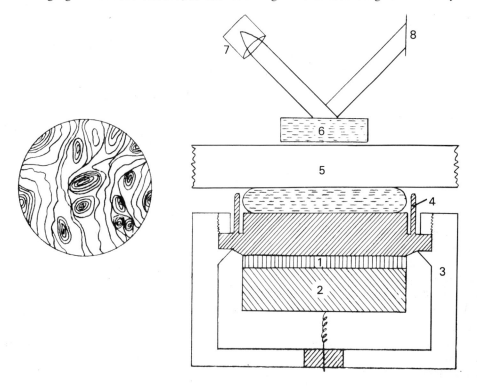

Fig. 1   The layout of the liquid surface levitation method invented by Sokolov to visualize sonic shadowgraphs of the objects of interest: 1, transducer; 2, backing of the transducer; 3, face plate; 4, mercury coupling; 5, material under investigation; 6, small container for liquid; 7, mercury lamp for illuminating; 8, screen.

technique called liquid surface levitation method were in essence visualized sonic shadowgraphs of the object of interest, e.g. a flaw. This method was then rediscovered in 1966 by R. K. Mueller and N. K. Sheridon[6] to form ultrasonic images by holographic techniques.

The next significant effort to produce ultrasonic images is linked with the name of Sokolov also, who created the first electronic acoustical-to-optical converter, an ultrasonic vidicon, in the early 1930s[7]. It was based on the observation that a piezoelectric plate yields an electric field replica of the acoustic field impinging on it, which when scanned by an electron beam can be visualized on the screen of a cathode ray tube. Based on this idea three types of ultrasonic vidicons have meanwhile been developed, one cathode-potential stabilized, another collector-potential stabilized, and an electron mirror type. Their practical application, nevertheless, is still very restricted.

13

The common characteristics of these and some similar techniques (e.g. different types of schlieren technique) for visualizing sound field lies in that the information carried by the sound waves is visualized indirectly and by multiplied transmission. N. Marinesco was first to demonstrate, as early as 1933, that there is a possibility of picture-like recording of sound field similar to photographs, when light waves act on light-sensitive materials. He demonstrated that even conventional photographic materials may be affected by ultrasonic radiation. To obtain acceptable sono pictures, however, he had to use considerable ultrasonic intensities and exposure times so that, unfortunately, people have been discouraged from doing further investigations in this direction. This idea was re-evaluated, and the method refined, only in the mid 1950s by Keck[8], by Arkhangelsky[9] and by Greguss[10]. The resolution of these sono sensitized photographic plates was rather good, so that the first acoustic hologram ever made[11] could be constructed on such a plate. The smallest necessary intensities to record a sonogram were in the order of 0.1 W/cm^2, exposure times varied between 30 and 280 seconds, but the technology associated with these methods was too time-consuming to use it in non-destructive testing or medical diagnostics.

Studies on the effect of ultrasonic waves on various dyes by A. R. Olson, N. B. Gordon[12], S. C. Liu, H. Wu[13], in 1932 led to the discovery of the so-called sonochrom substances, i.e. compounds whose color changes when insonified. Although some of them are rather sensitive and have quite a large dynamic range, no systematic studies unfortunately have been made yet on how they could be utilized in developing sono sensitive plates comparable to photographic plates. Although this endeavor is an unlikely area for a 'get rich quick' program, there would be a potential chance for an early reward.

When ultrasonic radiation acts on a suitable absorber, a spatial temperature pattern corresponding to the intensity distribution of the acoustic field is developed on its surface[14]. Coating an absorber with a layer of heat-sensitive material, such as cholesteric liquid crystals, the intensity distribution of the ultrasonic field can be visualized, i.e. a sonic image is formed, as demonstrated first by B. D. Cook and R. Werchan[15]. Although this method seems to be a simple technique, it has several serious shortcomings which discourage further investigations in this direction.

Another approach to the use of liquid crystals in acoustic imaging, proposed by L. W. Kessler and S. P. Sawyer in 1970[16], is to exploit their dynamic light-scattering properties. Unfortunately, ultrasonically induced light scattering is accompanied by formation of domain like regions which affect adversely the optical replica of the ultrasonic field.

The new liquid crystal acoustical-to-optical display was proposed in the early 1970s, based on the fact that ultrasonic radiation may affect the birefringent properties of a thin nematic liquid crystal layer[17]. Even ultrasonic holograms have been recorded by this technique.

This brief history on the attempts to form the optical replica of an acoustic field intensity distribution already shows that they are based either on sound—material interaction which results in an electrical signal proportional to the acoustic intensities,

14

and then this electrical signal is used to display the acoustic field in a visible form, (e.g. on the cathode ray tube), or on the interaction of the acoustic field with the recording material which results in the change of the optical properties of the material.

Recording techniques belonging to the first group generally use scanning methods, i.e. sampling procedures, those belonging to the second group are area detectors, i.e. they yield the ultrasonic information pattern directly in a two-dimensional form. At present, acoustic visualization methods based on sampling techniques are more advanced, but it would be wrong to think that nonsampling area detectors have no more future. To improve them, however, a better understanding of the interaction of ultrasonic waves with matter is needed. Incidentally, this holds true for the techniques belonging to the first group, too.

References

[1]GREGUSS, P. (1968) *Nature* **219** 482
[2]DREHER, J. J. (1969) *Acoustical Holography* Vol. 1 ed. A. F. Metherell pp. 127–138 Plenum Press, New York
[3]GREGUSS, P. (1971) *SPIE Seminar Proceedings* Vol. 24 pp. 55–83
[4]*Fr. Pat.* 505.703 (1918) P. LANGEVIN.
[5]SOKOLOV, S. Y. (1929) *Elektr. Nachr. Techn.* **6** 454–461
[6]MUELLER, R. K., SHERIDON, N. K. (1966) Appl. Phys. Letters **9** 328–330
[7]U. S. Pat. 2,164,125 (1937) S. Y. Sokolov
[8]MARINESCO, N., TRILLAT, J. J. (1933) *C. R.* Paris **196** 858
[9]KECK, G. (1956) *Acustica* **6** 543–545
[10]ARKHANGELSKII, M. E. (1960) *Akust. Zh. Akad. Nauk USSR* **6** 178–187
[11]GREGUSS, P. (1964) *Research Film* **5** 31–39
[12]GREGUSS, P. (1965) *Research Film* **5** 330–337
[13]OLSON, A. R., GARDEN, N. B. (1932) *J. Am. Chem. Soc.* **54** 791
[14]LIU, S. C., WU, H. (1932) *J. Am. Chem. Soc.* **54** 3617–3622
[15]ERNST, P., HOFFMAN, CH. W. (1952) *J. Acoust. Soc. Am.* **24** 87–88
[16]COOK, B. D., WERCHAN, R. (1969) 78th Meeting of the Acoustical Society of America, San Diego
[17]*Fr. Pat.* 2.177.410 (1973) P. Greguss

2 SONOGRAM ASSESSMENT BY INFORMATION THEORY

2.1 How good a visualized sonopicture can be

The basic theoretical principles involved in seeing by sound, i.e. the display of a two-dimensional optical replica of the insonified spatial scene can be understood without reference to the technology of any specific acoustical-to-optical conversion device, whether chemical, optical, or electronic mechanisms are involved in the conversion process. A reasonable basic specification for a device for seeing by sound might be that there should be a one-to-one relationship between the incident sound intensity and some measurable output state of the visualized sonic image. The quality of such a reproduction is generally judged by the user, having some *a priori* knowledge of the insonified scene, as 'that's a good (poor) likeness'.

This sort of judgement, however, is not at all reliable, especially in the case of sonic imaging, since there is a substantial difference between the interaction of the information carrier and the material under investigation, and that of the information carriers (generally light) on which the *a priori* knowledge of the observer is based. This difference arises because the scattering of the acoustic waves is dependent on the change in mechanical properties, whereas it is the change in the index of refraction that determines the scattering of electromagnetic waves. Therefore, objects which look transparent to optical radiation often show considerable acoustic contrast. This increase in contrast, when visualized, gives the observer a two-dimensional information pattern different from that to which he is accustomed. Sometimes it resembles, but only resembles, those structural details obtainable with chemical staining.

In optical imaging we can judge whether the picture which is obtained looks the same as the original piece of the real world it represents to the unaided eye. In sonic imaging, however, we have no subjective quality equivalent because we cannot see the sonic world directly. We never know really whether the visualized sonic image is sharp, clear and without distortion in the same sense as an optical image. Therefore, we have to establish objective standards. Unfortunately, the more recent scientific developments in optics, such criteria as resolving power and limiting resolution cannot be adapted directly to judge the quality of the visualized sonic image.

For instance, the two point resolution of Rayleigh criterion, which states that two equal luminous point sources are resolved when the centre of the diffraction pattern of the first falls into the first dark ring of the diffraction pattern of the second (Fig.

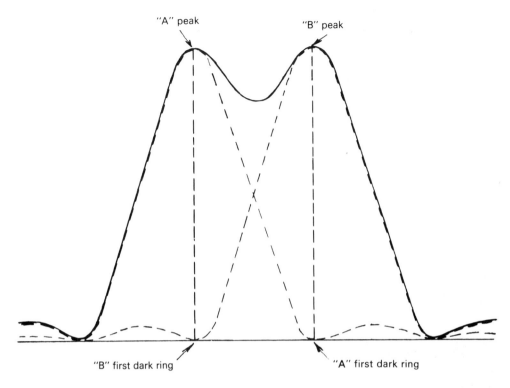

"A" peak

"B" peak

"B" first dark ring

"A" first dark ring

Fig. 2   Rayleigh resolution criterion.

2), has a meaning only on the optical side of the acoustical-to-optical conversion device, and it is practically meaningless on the acoustic side, since, in general, coherent acoustic illumination (insonification) is used. This is because with coherent radiation the phase of the two object points has to be taken into account. If we compare Fig. 2 with Fig. 3, it is immediately evident that the image will be identical to that under incoherent illumination only if the phase difference of the points separated by the Rayleigh distance is 90°, i.e. the terms of resolution holds only in this case. If two points have the same phase, they will not be resolved, and if their phase differs by 180°, they will be resolved much better than with incoherent illumination, the appearance of the sonic image will have a better 'quality'. The nebulous nature of the definition of quality depends also upon the intended use of the visualized acoustic image. To find, for example, a foreign body in an opaque eye one would like to have an image with sharp edges. Therefore, we judge in this case a sonic image with sharp edges as the 'best' image, and the fidelity of the rest of the image is immaterial.

In the ultrasonic differential diagnostics of soft tissues, however, we are interested in 'resolving' as many levels of shades of gray as possible, therefore, we shall judge

17

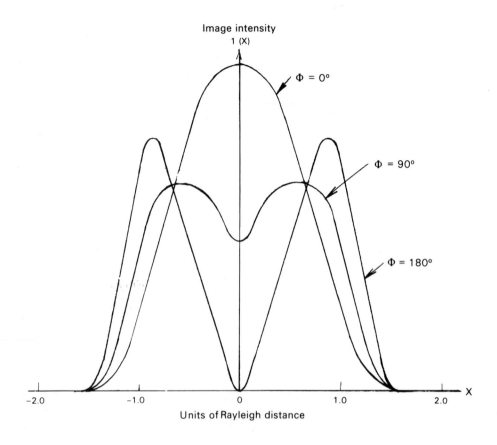

Image intensity
1 (X)

Φ = 0°

Φ = 90°

Φ = 180°

-2.0   -1.0   0   1.0   2.0

X

Units of Rayleigh distance

Fig. 3 Two point objects separated by one Rayleigh distance are variously imaged if coherently illuminated, depending on their phase relationship.

the quality of the sonic image with eight shades of gray to be 'better' than the same picture with two shades of gray.

It is interesting to note that shades of gray represent levels of $\sqrt{2}$ in luminescence of the image element, the pixel. This is, however, limited by the signal-to-noise ratio, e.g. two shades of gray need a dynamic range of S/N of $(\sqrt{2})^2$ or 2:1, while eight shades of gray call for $(\sqrt{2})^8$ which corresponds to an S/N dynamic range of 16:1. The problem of shades of gray-levels in acoustic imaging is, however, compounded by the additional question of coherent versus incoherent illumination.

The amplitude of the scattered coherent wavelets sent into the picture from the volume outside the focus region are summed *vectorially* and then squared. It is not the intensities that are summed as in the case of incoherent illumination, which gives an almost uniform background. This phenomenon called speckle noise results in intensity fluctuation which can be as high as 10 000:1. thus influencing adversely the

18

signal-to-noise ratio, i.e. the quality of the image will be judged to be less good than expected.

The complex amplitude distribution over the entire pupil of the acoustical-to-optical conversion system can be regarded as an intercepted part of information pattern flux conveyed by the acoustic waves scattered from the object, and the formation of the sonic image in the focal plane of the optical part of the conversion system can be considered as a decoding of this intercepted message, which represents the invisible pattern in a visible form. Therefore, the problem of judging a visualized sonic image is perhaps more tractable when treated as a problem of communication theory.

2.2 Shannon's theorem

The expression 'information' is used in communication theory in a very special way. To show what it really means, and to be in good style, we shall discuss it from the viewpoint of bats who have the natural capability of seeing images by sound.

Let us assume that a bat sees different sound-borne images from
a) 70 obstacles to be avoided (0)
b) 20 butterflies (B) and
c) 10 mosquitos (M).

For the bat the message of the sonic image of an obstacle to be avoided has the least information, since the bat can be very 'certain' that the echo forming the sonic image issues from a real obstacle to be avoided. On contrary, the message of the sonic image of a mosquito will have the highest information content since this message is what the bat least expects.

This example shows that in communication theory unexpected information is regarded having a higher value than a routine information. This interpretation attributes a numerical value to indirect information ratio to its probability. In our example, the probability of the first message for the bat is $p_0 = 0.1$, of the second $p_B = 0.2$, and for the third $p_M = 0.1$.

As early as 1928 Hartly[18], proposed for the information capacity of the communication channel the logarithm of the number of distinct symbols, n, capable for transmitting

$$\log n = \text{information capacity} \qquad\qquad 1$$

Shannon[19], later defined the information gain in the case of p symbols as

$$\log \frac{1}{p} = \text{information gain} \qquad\qquad 2$$

It is evident that when all symbols are equally probable i.e., P = 1/n, the information gained on receiving one of them will be

$$\log n = \log \frac{1}{p} \qquad\qquad\qquad 3$$

Unfortunately, this equality holds true only in special cases, and therefore log n is an inadequate measure of information. However, if the probability of a symbol occurrence is $p = 1/2$, and logs are taken to base 2, the associated information is then $\log_2 2 = 1$. This unit of information is called the 'bit' which is the abbreviation of binary digit, and indicates that a binary digit only conveys one bit of information when each binary state—a yes or no response—has *a priori* probability.

Information transmission is always accomplished in a thermodynamic system having a temperature of T_0 and a minimum noise output of $Z = k\, T_0 \triangle f$, where k = 1.38 x 10^{-23} J/^{0}K being the Boltzmann constant, and $\triangle f$ the bandwidth. Marko[20] has demonstrated that the energy necessary to transmit one bit of information at room temperature is 3 x 10^{-21} Ws, i.e. transmission of information is an energy-consuming process. In other words information is necessarily attached to a certain amount of energy, more correctly, to an energy package. Therefore, transmission of information means transmission of energy, and from the point of view of image assessment it is not of interest what type of energy it is, electromagnetic (light), mechanical (acoustic), etc.

From the examples cited it follows that at a conceptual level, information theory and the definition of thermodynamic entropy have a common origin, and differ only by choice of units. However, it should be kept in mind that the 'bit' as scale factor for information has its validity only if both answers—yes-or-no—have equal *a priori* probability.

2.2.1 The influence of noise

The received message, however, will not be necessarily identical with that originating from the reflecting target since during transmission environmental and communicational elements usually introduce noise to the message. But it has to be kept in mind that there is difference between distortion and noise. The latter one is retained for those effects where the signal does not undergo the same predictable change during transmission, while in the case of distortion a particular transmitted signal *always* produces the same received signal. The incorrect interpretation of this distortion can, however, be avoided by performing the inverse process as the received message, if no two different transmitted signals produce the same received message, i.e. if a distortion had an inverse.

According to this discrimination, when noise is present—which means a chance variable governed by the statistical process at work during transmission—it is very difficult, if not impossible, to reconstruct with certainty the original message from the received signal. Shannon[21], and others[22], have shown that the amount of the missing information can be measured by the entropies associated with signal and noise statistics. It stands to reason that when there is no noise the entropy of the input, $H(x)$, and the entropy of the output, $H(y)$, is equal, while when noise is present a conditional entropy is attached to output $H_x(y)$ or to the input $H_y(x)$ depending on whether the input or the output is known. The joint entropy may be defined as

$$H(x,y) = H(x) + H_x(y) = H(y) + H_y(x) \qquad\qquad 4$$

so that the rating of transmission will be the amount of information sent minus the missing information:

$$R = H(x) - H_y(x) \qquad\qquad 5$$

or

$$R = H(y) - H_x(y) \qquad\qquad 6$$

The average ambiguity in the received message is measured by the conditional information sent, which is called equivocation. If the noise is additive and independent of the signal, equivocation will be a function of only the difference $z = y - x$, and so the information rate can be expressed as the entropy of the received signal minus the entropy of the noise:

$$R = H(y) - H(z) \qquad\qquad 7$$

Shannon[21], also demonstrated that if Z is the average noise power, binary digits can be transmitted with appropriate coding, at a rate of

$$H(y) - H(z) = \Delta f \log_2 \frac{P+Z}{Z} \text{ bits/sec} \qquad\qquad 8$$

with an arbitrarily small frequency of errors, provided that the noise is white Gaussian. Solutions of other types of noise statistics and other constraints are not so straightforward but, fortunately, they are also not so important either in the first steps of investigating the problem of seeing by sound. Those who are interested in further details may be referred beside Shannon's original papers to the works of Bartletson and Witzel[23], Brillouin[24], Gabor[25], Tribus and McIrvine[26], and Woodward[27].

21

2.2.2   Relevance of sound images

Let us assume that a sonic image (not the visualized sonic image) consists of a set of individual pixels, each of area A. Each of these pixels can be regarded as storage units in which the amplitude-bound information is stored in the form of a number. Since all physiological and man-made receptors are sensitive only to the square of the amplitude of the information-bearing wave[28], these numbers are in reality intensity values which, when visualized, manifest themselves as density levels. Therefore, following the train of thoughts of Danity and Shaw[29], concerning the photographic application of information theory, we can gain some insight into the problems of seeing by sound.

Sonic images are formed practically always in noisy environment, and so the intensity value manifesting itself as density level in a pixel is subjected to uncertainty. In order to be able to evaluate these density levels, i.e. to 'perceive' the sonic image, the levels must be separated by a sufficient density interval prescribed by the evaluating system.

If σ_A denotes the mean-square sonic image density fluctuation as measured with a scanning aperture equal to the area of a pixel A, and 2k represents the standard deviation by which adjacent levels are separated, it can be shown[30], that the number of possible recording levels is as follows:

$$M = \frac{R}{2k\,\sigma_A} + 1 \qquad\qquad 9$$

where R represents the total density range of the sonic information pattern.

Assuming further that the pixel size A is in the same order or greater than the low frequency value of the Wiener spectrum of the noise, i.e.

$$G = A\,\sigma^2_A \qquad\qquad 10$$

then

$$M = \frac{RA^{1/2}}{2kG^{1/2}} + 1 \qquad\qquad 11$$

The information capacity will be $\log_2 M$ bit/pixel, provided each of the recording levels is equiprobable. Since there are N cells per unit area, the information per unit area will be

$$C = N \log_2 M \qquad\qquad 12$$

Combining Equation 11 and Equation 12, and taking into account that $N = A^{-1}$, the information capacity per unit area can be written in the form of

22

$$C = \frac{1}{A} \log_2 \left(\frac{RA^{1/2}}{2kG^{1/2}} + 1 \right)$$

13

As a consequence of information theory at least two recording levels are needed, and the area of a pixel which allows binary coding can be defined from Equation **13** as

$$A = \frac{2k}{R} G$$

14

In order to be able to estimate the value of the criteria for the separation of recording levels, k, we have to observe the capability of the readout system, e.g. that of the brain of the bat, or that of the acoustical-to-optical conversion display in our acoustic imaging device.

Since in general we may assume that image noise fluctuations are normally distributed, allowing an error rate of about 1 in 10^6, a separation of $\pm \sigma_A$ would be enough according to Altman and Zweig[31]. To be on the safe side, we would suggest $\pm 10 \sigma_A$, and so from Equation **14** for binary coding, i.e. when the density range is $R = 2$,

$$A_2 = 100 \ G$$

Using Equation **13**, the minimum possible pixel size for recording 3, 4 etc. levels would be

$$A_3 = 400 \ G, \ A_4 = 900 \ G, \ \text{etc.}$$

It is thought[32], that an acoustic imaging system will have a good performance if the pixels are, in terms of the information-bearing sound wavelength λ, not greater than $(5/2) \lambda^2$. Assuming a frequency of 3 MHz and an average sound velocity of 1.5 x 10^5 cm sec^{-1} (average sound velocity in biological tissues), the ideal pixel size would be

$$C = \frac{1}{6.25 \times 10^{-4}} \ \log_2 2 = 1.7 \times 10^{-2} \ \text{bit cm}^{-2}$$

in the case of 3-level recording.

$$C = \frac{1}{6.25 \times 10^{-4}} \ \log_2 3 = 2.6 \times 10^2 \ \text{bit cm}^{-2}$$

and for 6-level recording

$$C = \frac{1}{6.25 \times 10^{-4}} \ \log_2 6 = 4.8 \times 10^2 \ \text{bit cm}^{-2}$$

23

Comparing the capacity of the binary and 3 or 6-level recordings it has to be realized that multilevel recording may offer little practical advantage over the binary case, and may result in coding complications and difficulties. This conclusion follows not only from Equation **12**, but was also confirmed by the experimental results of Altman and Zweig[33]. More can be achieved if the size of the pixel can be reduced, i.e. the number of the pixels per unit area increased. There may, of course, be practical difficulties in this regard too, since the pixel size is defined as the area for which the point spread function is reduced to, say, 10% of its peak value, but these problems are less severe in general.

The maximum information capacity of a sonic image has been expressed in terms of bits per unit image area so far, but no account has been taken of any situation of read-in information. The necessity for developing a practical real-time acoustic imaging system, e.g. for diagnostics, requires also the evaluation of this problem. This will be discussed in a later chapter.

The simple analysis and the examples presented have demonstrated that an acoustic information pattern can only be coded in an optimum mode as a sonic image if it is done according to entropy considerations. It must be kept in mind that *information capacity* and *information content* are two different things. The first one is a measure of optimum coding, where distinction has to be made between a binary digit and a bit, while information content is related to the results achieved under practical conditions. This, however, means that practical coding of sonic images is strongly related to the materials and methods used by different recording techniques and will therefore be discussed in that chapter.

The usable information content of a displayed sound image depends, however, also on the human visual system, and therefore this has to be taken into account when performance specification is given. So it has to be kept in mind that, for instance, the contrast which the insonified and visualized target must have before it can be discerned against the background decreases with increasing target size and with increasing background luminance, or that for visualized sound images which are presented for a shorter 'exposure time' than the integration time of about 0.1 sec described by the Bunsen-Roscoe law, more contrast is needed.

If the insonified target to be visualized moves, the situation becomes far more complicated. The fovea of the eye integrates all the radiation it receives at each point within a time interval of about 0.1 sec if it does not move. However, when there is a relative motion between the visualized sound image and the eye, this exposure time is determined by the angular rate of motion and the target's visual angle in the direction of motion as

$$\text{Exposure time} = \frac{\text{target angular size}}{\text{angular rate}}$$

The Bunsen-Roscoe law applies, therefore, for moving targets only if this expression is used for exposure time.

References

[18]HARTLEY, R. V. L. (1928) *Bell Syst. Tech. J.* **7** 535–550

[19]SHANNON, C. E. (1948) *Bell Syst. Tech. J.* **27** 379–423, 623–653

[20]MARKO, H. (1966) *Kybernetik* **3** 128–136

[21]SHANNON, C. E., WEAVER, W. (1949) *The Mathematical Theory of Communication* University of Illonoi's Press, Urbana

[22]BRILLOUIN, L. (1961) *Science and Information Theory* Academic Press, New York, London

[23]BARTLETSON, C. J., WITZEL, R. F. (1967) *Photogr. Sci. Eng.* **11** 263–268

[24]BRILLOUIN, L. (1961) *Science and Information Theory* Academic Press, New York, London

[25]GABOR, D. (1961) *Progress in Optics* **1** 109–119

[26]TRIBUS, M., MCIRVINE, E. C. (1971) *Scient. Amer.* **224** No. 3 179–

[27]WOODWARD, P. M. (1953) *Probability and Information Theory with Applications to Radar* Pergamon Press, Oxford

[28]GABOR, D. (1948) *Nature* **161** 777–779

[29]DAINTY, J. C., SHAW, R. (1974) *Image Science* Academic Press, London, New York

[30]LEVI, L. (1958) *J. Opt. Soc. Am.* **48** 9–16

[31]ALTMAN, J. H., ZWEIG, H. J. (1963) *Photogr. Sci. Eng.* **7** 173–178

[32]WANG, K. Y., WADE, G. (1974) *J. Acoust. Soc. Am.* **56** 922–928

[33]ALTMAN, J. H., ZWEIG, H. J. (1963) *Photogr. Sci. Eng.* **7** 173–178

3 SOUND AS AN INFORMATION CARRIER

The word 'sound' covers two conceptions: a physical and a physiological one. Physically all mechanical vibrations are sounds since there is no principal difference in physical detection between the two conditions, while physiologically, only mechanical vibrations which are in the range of normal perception to human ear are considered as sound. In both cases, physical properties of the medium, for instance density, demonstrate periodic changes depending on the space coordinates, and these changes propagate with a finite velocity. For standing waves, this change of property is dependent on the space coordinates, and for travelling waves the change is also time dependent. The degree of change is proportional to the energy which provokes this change. What is actually transmitted is a perturbation of some sort in the equilibrium arrangement of the atoms of the material, since material in some form is always required for the transmission of these waves, in contrary to the propagation of electromagnetic waves, i.e. light. It has to be emphasized, however, that only the disturbance is propagated, and the material itself does not go anywhere.

3.1. Sound characteristics

Although light and sound differ in their propagation medium, the fundamental wave propagation characteristics are very similar, nearly identical. The term *particle*, however, in the description of sound propagation is to be understood as a volume element large enough to contain millions of molecules, so that it can be regarded as continuous, but small enough so that its material characteristics can be regarded as constant.

One may obtain a physical conception of sound waves by considering that any given instant the particles of the material in which the sound is propagating, will have a displacement to one direction and, then, to the opposite. The velocity, V, of a vibrating particle is given by

$$V = \omega A \cos \omega t \tag{15}$$

where A is the maximum displacement of the particle, the amplitude of the vibration (ωA is sometimes called velocity amplitude) and $\omega = 2\pi N$, N being the *frequency* of the sound wave, which is the measure of numbers of waves that pass a particular point per second, and is measured in Hertz (Hz) = 1 cycle per second.

At ultrasonic frequencies this displacement becomes very small, being 10^{-9} m at 1 MHz in a steel block for power density of 1 W/m^2. This approaches atomic dimensions and is an essentially important characteristic in conveying information useful in material inspection.

Particle displacement is also a good example for the difference between the propagation of light and acoustic waves. As long as frequency, wavelength, wave velocity have their counterpart in electromagnetic wave theory, particle displacement has no clear counterpart.

The distance between adjacent regions of *equal positions* of the particles is called wavelength, λ, and since it is desirable to have some method to specify whether a wave is at its peak, is approaching it, or is at some other point in the cycle, the idea of *phase* has been introduced. Phase is measured in degrees or radians, and indicates in what part of its vibrational cycle the wave happens to be at a particular instant of time.

According to the definition of frequency, the wave motion covers $N\lambda$ distance in unit time i.e. the sound velocity,

$$c = N\lambda \qquad\qquad \textbf{16}$$

or for the wavelength,

$$\lambda = \frac{c}{N} \qquad\qquad \textbf{17}$$

Since complete information on the types of acoustic waves—the term 'acoustic' is applied to all mechanical vibrations, similarly, the terms 'sound' and 'sonic' are not restricted to audio range—and as their propagation characteristics are readily available in textbooks[34, 35, 36] or in encyclopedias[37], we shall deal with them only so far as it is necessary for the understanding of how seeing by sound can be accomplished and how it may help to improve our knowledge of nature.

3.2   Types of waves

Longitudinal waves are the most common acoustic waves. It is characteristic of these waves that the motion of the particles in the medium is parallel to the direction of wave propagation. It should be pointed out here that the longitudinal wave is not to be confused with the total vibration of a part in which the sound wave is travelling. Such vibrations are characteristic of the geometry of the part and are caused by the action of the wave.

Longitudinal waves can be propagated in solids, liquids or gases. When travelling in solids, the compressional wave causes not only temporary changes in density, but shear forces which are perpendicular to the direction of propagation are also pro-

voked. In finite solids these resulting shear or transverse waves, have particle movement, in the medium which is at right angles to the direction of wave propagation.
Their velocity in a metal is approximately one-half of that of the longitudinal waves,
which means that the wavelength of shear waves is approximately one half of that of
the longitudinal waves of the same frequency. Since shear forces do not exist in
gases, and as a first approximation they can be neglected in liquids too, shear waves
can be generated significantly only in solids.

When sound waves are propagating on a solid surface in air the particles on the
surface move in longitudinal as well as in transverse direction, and as a result, the
particle orbit is an ellipse. These waves are called surface or Rayleigh waves. Their
basic property is that if a solid is at least several wavelengths thick, the displacement
of the particles a few wavelengths below the surface is negligible. The wave velocity
depends upon the material itself and is about nine-tenths of the shear wave velocity.

Lamb or plate waves are another very important type of wave and their propagation
characteristics bear a lot of similarity with the propagation of electromagnetic waves
in wave-guides. As their name indicates, they arise only in thin (i.e. compared with
the wavelength) solid materials such as plates and rods. These types of sound waves
can play an important role in recording acoustic images and their nature is demonstrated in Fig. 4. The different types of plate waves propagating in an aluminium
plate as functions of wall thickness and frequency is shown.

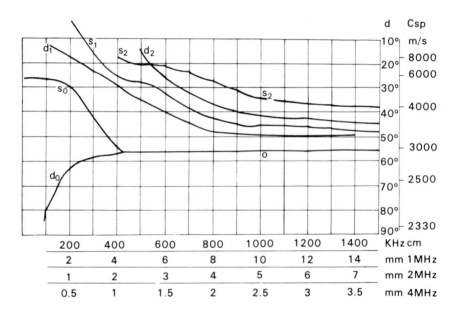

Fig. 4 Different types of plate waves propagating in an aluminum plate as function of wall thickness and
frequency.

28

3.3 Nature of a sound field

The propagation of sound waves is always accompanied by an alternating pressure which can be characterized as

$$P = P_o + ZA \, \omega \cos \omega t \qquad\qquad 18$$

where P_o is the hydrostatic pressure, A is the displacement amplitude and Z is the *acoustic impedance* of the medium. $\omega/2\pi$ is the frequency of the sound wave. It should be noted that $ZA\omega$ is the pressure amplitude of this sound wave. Z is a complex quantity and can be expressed as

$$Z = Z_o + jX \qquad\qquad 19$$

where jX is the reactive component arising from the inertia and stiffness of the medium. For a plane wave the characteristic impedance (also called specific impedance or wave impedance) is given by $Z_o = \rho c$ where ρ is the density of the medium and c is the sound wave velocity. It should be noted that the expression $AZ\omega$ is sometimes called alternating pressure amplitude, or pressure amplitude.

When a travelling wave acts upon an obstacle, beside the alternating pressure a unidirectional pressure, the so-called sound radiation pressure is also effective. It can be shown that the sound radiation pressure is equal to the mean energy density of the travelling wave, and can be given by

$$E = 2\pi^2 \, \rho N^2 A^2 \qquad\qquad 20$$

where N is the frequency of the wave of amplitude A and ρ is the density of the medium.

Since the intensity of the acoustic radiation is equal to the energy flowing through unit area in unit time in the form of a plane wave, it can be written that

$$I = Ec = 1/2 \, (c\rho\omega^2 A^2) \qquad\qquad 21$$

Now the particle velocity amplitude
$$\omega A = \frac{P}{\rho c}$$
$$= \frac{P}{Z_o}$$

(Ohm's law of acoustics) so Equation 21 becomes

$$I = P^2/2\rho c = P^2/2Z_o \qquad\qquad 22$$

The unit of acoustic intensity is 1 $erg/cm^2 sec$, and the practical unit is $10^7 \, erg/cm^2 sec = 1 \, watt/cm^2$.

The decibel (dB) is sometimes quoted as a measure of sound intensity, but it must be emphasized that dB, being the logarithmic ratio of two intensity levels, is always a relative measure. Therefore, the intensity level IL is given by

$$IL = 10 \log I/I_o \text{ dB}$$

where $I_o = 10^{-16}$ W/cm^2 is the reference level. If the intensity of a sound field is higher than 10^{-3} W/cm^2—especially in gases and liquids—acoustic radiation may be the source of permanent material changes. Therefore, sound waves with larger intensities are often called 'supersonics', irrespective of their frequencies.

3.4 Laws of propagation

For the transmission of sound a material medium is necessary and the energy of excitation usually comes from a vibrating body in contact with the medium or the agency of excitation could be thermal or electromagnetic radiation impinging on the medium.

If the sound source is a point source, the sound waves spread evenly in all directions (assuming isotropic medium), and, according to Huygens' principle, we have a purely spherical wave. At a sufficiently great distance from the source, however, a small part of the sphere can be considered as plane, i.e. we have a pencil of rays as a parallel beam giving a plane wave. Consequently, the 'sufficiently great distance' or what is equivalent to this, a high ratio of the dimensions (D) of the sound source to the sound wavelength (λ) generated determines whether we can consider a sound wave as a plane wave or not. In the audible range, this condition rarely holds, since the dimensions of the sound source are always smaller or at least in the same order of magnitude as the wavelength, but in the ultrasonic range, this condition is satisfied in most cases. This means that the laws of geometrical optics can be adopted for evaluating the path of the ultrasonic beam, if certain restrictions are taken into account.

The ultrasonic beam which emerges from a plane surface is considered as a parallel beam until a distance L, which is called near-field or Fresnel zone (Fig. 5). This near-field may be regarded as a cylinder with diameter D of the vibrating body. The relation between D and L is given by

$$L = \frac{D^2}{4\lambda} \qquad\qquad 23$$

In this near field, sound intensity varies in a manner which is very difficult to follow and to control. When, however, sound is being used as an information carrier, information is bound to intensity, i.e. to the square of the amplitude, and therefore it may sometimes be desirable to know the relative intensity distribution in the sound field of the transducer. A quite good description of the sound field can be given by

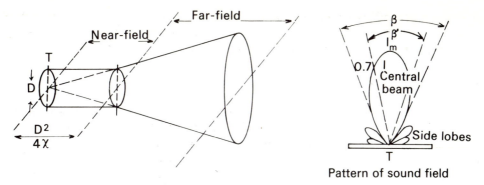

Fig. 5   Approximate form of ultrasonic beam for large values of D/λ. T is circular piston transducer of diameter D.

using the following equation for the axial intensity distribution in the near-field:

$$\frac{I_y}{I_{max}} = \sin^2 \frac{k}{-2}\left(\sqrt{\frac{D^2}{4} + y^2} - y\right)$$

24

where I_{max} is the maximum intensity on the axis of the transducer, I_y is the intensity at axial distance y from the transducer, and k is the wave number.

At the end of the near-field the parallel beam becomes more and more divergent and reaches the so-called far field or Fraunhofer zone. Here, intensity variations due to interference effects gradually disappear and the beam spreads, with approximately uniform intensity over the wavefront. This spread is a function of the ratio λ/D and can be given by

$$\sin \beta \approx 1{,}22 \; \lambda/D$$

25

where β is the half solid angle of the beam divergence.

3.4.1   Acoustic waves at boundaries

When the acoustic energy reaches the boundary of a new medium—supposing that it is perpendicular to the direction of propagation—one part penetrates into the new medium while the other is reflected back. The ratio of these two parts depends upon the specific acoustic impedance, Z. If the specific acoustic impedance of the two media are equal, all the acoustic energy penetrates into the second medium. If not, the reflected energy is given by

$$A_r = \left(\frac{m-1}{m+1}\right)^2$$

26

where $m = Z_1/Z_2$, (subscripts 1 and 2 refer to the two media). For penetration, however,

31

$$A_p = 1 - A_r = \frac{4m}{(m+1)^2}$$

The values of Equation **26** and Equation **27** do not change if $1/m$ is written for m, i.e. the reflected energy remains the same when the acoustic energy is travelling in the opposite direction.

When the density of and the sound velocity in the media are known the reflection coefficients can easily be evaluated as is shown in Fig. 6. If, however, the thickness

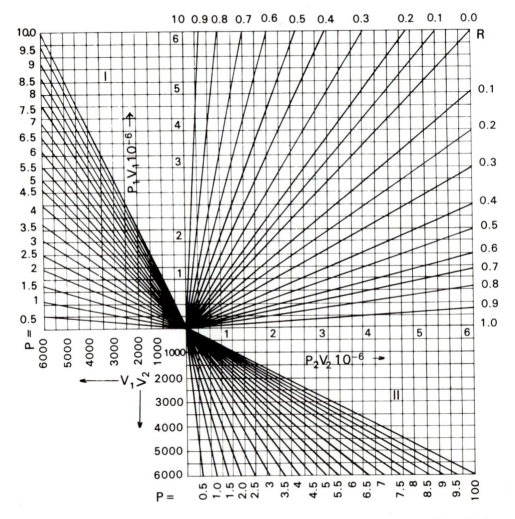

Fig. 6 Evaluation of reflection coefficient. A_o = first amplitude, A_r = reflected amplitude, R' = reflection coefficient.

of the medium in which the acoustic energy penetrates is not larger by at least one order of magnitude than the sound wavelength, i.e. it is not comparable with the wavelength, the situation becomes more complicated as is shown in Fig. 7.

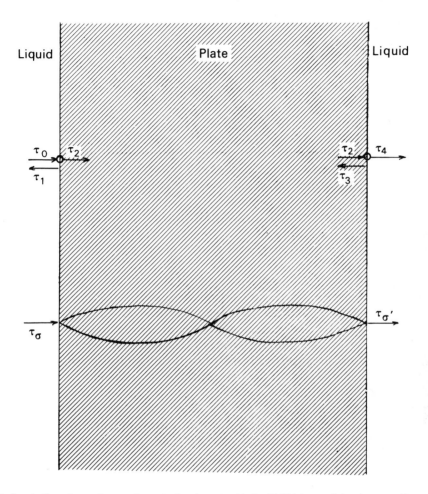

Fig. 7   Penetration of sound waves through plate immersed in liquid. Thickness of the plate = n.λ/2.

Let us assume that a solid plate of $n\lambda/2$ thickness is immersed in a liquid. When the acoustic wave reaches the boundary, according to Equation **27** one part of it penetrates into the plate, and the other part is reflected from the second surface in accordance with Equation **26.** The reflected wave, however, intersects at a favorable position with the following wave so that standing waves are built up in the plate. Therefore, the plate vibrates with larger amplitude, and so more acoustic energy passes through the plate. The reflection coefficient in this case can be given by

33

$$A_r = \frac{(m^2 - 1)^2}{4m^2 \; ctg^2 \; (\frac{2\pi d}{\lambda_r}) + (m^2 + 1)^2}$$ **28**

where, d is the thickness of the layer, and λ_r is the wavelength in the layer.

In the above example the denominator is infinite, consequently there is no reflection and the total acoustic energy penetrates through the layer. When, however,

$$ctg \; \frac{2\pi d}{\lambda_r} = 0$$

λ_r is maximum, i.e. most of the acoustic energy is reflected. It can easily be shown that this is the case when the thickness of the layer is an odd multiple of $\lambda/4$.

The other interface characteristic, reflection, which occurs when the incident of the wave is not equal to 90°, depends only upon the propagation velocities in the two media, the ray path being in accordance with Snell's law

$$\frac{sin\alpha}{sin\beta} = \frac{c_1}{c_2}$$ **29**

where α is the approaching angle with respect to the interface normal, and β the departing angle.

For transmission between a medium of lesser velocity into a medium of greater velocity a critical incident angle exists for which the reflected angle is 90°. No energy can be transmitted into the second medium if the incident angle exceeds the critical angle: in this case all energy will be reflected.

Longitudinal waves are transmitted through gases and liquids, but both longitudinal and shear waves will propagate through solids. Thus, an acoustic wave entering the solid can be split into two rays which propagate at different velocities and reflection angles.

For the sake of simplicity we do not discuss here the energy distribution between the various reflected and refracted waves in the case of oblique incidence, it can be found in the literature[38], we only stress that the energy distributed to these various wavefronts must total the incident energy.

3.5   Absorption

A portion of propagating acoustic energy may be converted into heat. It results from the nonelastic compression of the medium described by a reactive loss term added to compressibility.

Intensity loss after the wave has travelled a distance x is given by

$$I_x = I_o \; exp(-2\alpha x)$$ **30**

where α is the amplitude absorption coefficient and has the dimension of cm^{-1}.

Since the relation between intensity and amplitude A is quadratic(**21**),

$$I = ZA^2\omega^2/2,\qquad\qquad\qquad\qquad\qquad\textbf{31}$$

then the amplitude A_o will drop to a value of A_x after the wave has travelled distance x, accordingly

$$A_x = A_o \exp(-\alpha x)\qquad\qquad\qquad\qquad\textbf{32}$$

It has to be emphasized that the intensity absorption coefficient α is always two times larger than the amplitude absorption coefficient. Unfortunately, this is generally not emphasized in the literature and often leads to misunderstanding. The absorption coefficient, however, is not frequency independent. It varies according to the equation

$$I_x = I_o \exp(-2N^2\alpha'x)\qquad\qquad\qquad\qquad\textbf{33}$$

where α' is the frequency independent absorption coefficient. Sometimes it is more convenient to use the dimensionless absorption coefficient, α^* which is the absorption over a wavelength. In this case

$$I_x = I_o \exp(-2\alpha^* x/\lambda)\qquad\qquad\qquad\qquad\textbf{34}$$

It is the general convention to use the value of the absorption coefficient α not in terms of cm^{-1}, but in terms of neper/cm. Its value is given by the ratio of the two quantities, which is equal to the natural logarithm e = 2.178. This means that the value of $1/\alpha$ is indicative of the distance (in cm) during which amplitude A_o of the vibration diminishes to A_o/e. So,

$$\alpha^* = \alpha\lambda = \lambda N^2 \alpha'$$

Occasionally the absorption coefficient is given in dB/m in literature. Since, 1 neper is 8.686 dB, the value given in neper/cm has always to be multiplied by 8.686 if calculations are to be made in dB/m.

3.6 Information is bound to . . .

As we have seen, information (more correctly, the signal representing information) is always attached to an energy package. To answer the question which attribute of a wave may carry information on consistence, storage or position in space when it is attached to mechanical energy, we have to investigate further the characteristics of

a) frequency and/or wavelength,
b) amplitude—intensity, and
c) phase.

A *single frequency* cannot carry information on the shape and on the range in space of the object, since any point in space could be the origin of waves of any frequency. In optical terms, red light shows an object to be of the same shape as when illuminated with green or yellow light.

Multifrequency package (broadband impulse), however, can carry information on shape.

Varying frequency can carry information on object motion.

Amplitude (intensity) alone cannot carry information on distance. In optical terms, objects in the same plane and same distance from the observer, even though illuminated by different intensities, still appear to be at the same distance.

Amplitude ratio, however, can carry information on the consistency of the material through which the wave has travelled.

Amplitude distribution of a wavefront can carry information on shape. Such an information pattern is called an image. The simplest image is a shadow.

Phase, since it indicates in what part of the vibration cycle the wave happens to be at a particular instant of time, can carry information on range (distance).

Thus, to describe a 3-D space or a part of it, i.e. an object, both the amplitude and the phase of the information bearing wave have to be processed. This is a task much more complicated than one would have thought. A monochromatic, noncoherent beam carries with it n degrees of freedom. But the information on the beam is not a vector in n dimensions, as one might think, but a tensor with n^2 dimensions[39]. One could in principle determine every one of these by experiments in two planes, but—as pointed out by Gabor[40]—the method is hopelessly complicated. Fortunately, the situation is somewhat more simple if the information carrier is coherent radiation, because in this case only 2n data are required: only a well defined amplitude and phase are related to every point. Nevertheless, the simultaneous recording and/or reconstruction of the information carried by the amplitude and phase of the wavefront causes problems, since the energy sensors are always square law detectors which do not record phase information.

This problem was then solved by Gabor in 1947, when he invented holography[46], and after having realized that acoustic waves, especially ultrasonic waves are highly coherent, the idea has been extended to mechanical waves by Greguss[42] in 1964/65. One may, however, ponder whether extending holography to acoustic waves as information carriers is really needed to see by sound. Since, except for ultrasonic wavelengths approaching that of the visible light, i.e. in acoustic microscopy, there are linear detectors available and so both amplitude and phase information of a sound wave can be processed.

3.6.1 Perception of sonic images

There are two major sets of factors governing the perception of man and his devices forming sonic images. One set of factors relates to the physics of sound, to the acoustic equivalent of geometric optics (there is no separate term for it yet) and the engineering approaches to the design of acoustical-to-optical converters. The second set is related to factors generally disregarded when sonic images are evaluated, such as laws of psychophysics and vision, the interactions among acoustical (ultrasonic) tasks, the time available for doing it, and other subjective matters related to the observer and his task.

An overall representation of the process of forming and perceiving sonic images from insonification, the acoustical equivalent of illumination, to final decision is shown in Fig. 8. Problems related to the second set of factors are biased by the first

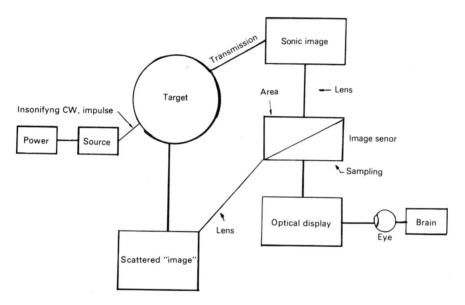

Fig. 8   Flow diagram of the process of forming and perceiving sonic images from insonification to final decision making.

set of factors, the effect of which are then summarized at the display. This really means that the perception of sonic images is strongly related to, and is a function of the perception of the displayed sonic image information. Little if any work has yet been carried out to establish the link between the objective measures of the displayed sonic image and the subjective measures of the sonic image quality which is strongly related to the application, i.e. to the character of the insonified scene, its spatial spectral composition, its contrast, the radiance of the display, etc., and to the difficulty and degree of details of the visual task to be performed on the basis of sonic image. Further, the amount of time the observer has to make his observation including the details is usually overlooked in discussing the performance of these displays. This is all the more regrettable since a great number of excellent publications exist on the perception of displayed information other than sonic images that could be applied to the problems of seeing by sound. A complete survey of this subject can be found in the book *Perception of Displayed Information* edited by L. M. Biberman[43].

Our objective is not to discuss these parameters in detail, only to review in this sense the methods and techniques of acoustical-to-optical conversion. Nevertheless, we wish to show on a single example how important these questions can be in designing acoustical-to-optical displays.

Sonic image quality and even the quality of the displayed sonic image is not a single prerequisite for good sonic image perception. The displayed sonic image must be large and bright enough to perceive in the right way the acoustic message converted into a luminous message for the retina. However, large acoustical-to-optical displays are not always necessary. Large *apparent* display may be used, providing the magnifying technique does not appreciably degrade the overall modulation transfer function (MTF) of the system.

Generally, displays are viewed from about 50 cm, the observer therefore needs an image size of about 2.5 mm. This indicates that the observer will usually not see anything on the display that is smaller than $\sim$ 2.5 mm. If the size of the display is $\sim$ 10 cm, there are only about 40 possible 2.5 mm spots across the display, and it is going to be one of the these 40 spots that will attract the observer's attention. It makes therefore no sense to use, for example, a scanning acoustical-to-optical conversion method which 'feeds' the display with more than 40 lines.

3.6.2 *Sound transmission and spatial information*

The flow diagram of Fig. 8 shows clearly that sonic images—in the sense that an image is *any* two-dimensional projection of wave energy conveying information as a variation in wave intensity—can be formed either by transmission or by reflection. If, however, the term 'image' means a two-dimensional intensity pattern which has a spatial point-by-point correspondence to the original insonified two-dimensional scene, i.e. it is a more or less faithful reproduction of the object, then the images obtained by transmission technique differ basically from those obtained by reflection. It must be emphasized further that the two dimensions represented are usually the vertical, or altitude, and the horizontal, or azimuth. If the original scene is a three-dimensional one, i.e. it contains range information too, only its *projection* is pictured in the sonic image, provided no holographic methods are applied. This is true if either continuous waves (CW) or impulse technique is used.

When sound is transmitted through a target, only that part of sound energy which has not been scattered will produce a sonic image in the sense of an optical image. The intensity distribution of the sonic image conveys information on the apparent average sound two-dimensional absorption property of the insonified target. Apparent means that the gray scale in the sonic image is not only the result of the spatial distribution of the absorption coefficient in the target, but of the acoustic impedance variation too, and the two effects cannot be separated.

Shadow casting is a special case of sonic transmission image formation, and can be regarded as a two-level recording. Although it is a simple imaging technique, it has its drawback. A sharp shadow is cast only immediately in the front of the surface, then, due to the diffraction of the information bearing wave, spreads and diminishes in intensity rapidly. Even for a target circumference of 18 times the wavelength, shadow strength will have diminished to 42% at 10 diameter distance, as is shown in Fig.9.

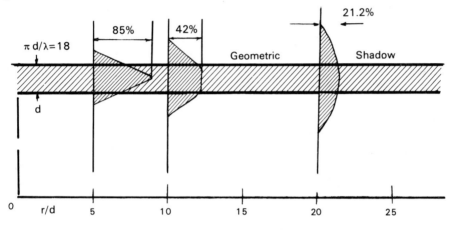

Fig. 9   Problems of shadow casting.

It was the combination of these factors that negated all efforts to use transmission imaging when sonic imaging is considered to solve a technical or medical problem. This relatively simple sonic imaging method, however, now shows a 'come-back', as we shall see in Chapter 6.4.

3.6.3   Sound reflection and spatial information

In contrast to the transmission sonic image, the sonic image obtained by reflection is never a faithful reproduction of the target, although it can be regarded as the antithesis of the transmission mode. The reason for this is that the two-dimensional pattern is the result of randomly scattered rays and so spatial information in altitude and azimuth are not contained in a direct form. Only if mirrors or lenses are used to reorganize (to focus) the scattered wave energy can a faithful reproduction of the insonified target be achieved.

Sonic images similar to those resulting from shadow casting can, however, be obtained if the scattered field is scanned in two dimensions with an aperture having an area equal to the image pixel, and is oriented so that it is always at normal incidence to the propagating part of the reflected sound energy. Such a sound image is called C-mode.

If the received energy is quantized, the appearance of the sonic image will be similar to that of a transmission sonic image; nevertheless, its information content will be different, depending upon the reflecting surface characteristics and especially its smoothness with respect to the wavelength of the sound energy used.

So when the reflecting surface is irregular in shape with respect to the wavelength of the insonifying beam, it will strike the surface at different angles of incidence so that the reflected energy will be scattered widely and only a part of it will reach the scanning aperture at normal incidence to its propagation path. On the other hand, when the reflecting surface is smooth, there will be only a single deviation in which the beam is reflected. Such reflections are called specular.

In summary, transmission sonic images give information mainly on those properties of the insonified target which are associated with sound absorption, while reflection sonic images mainly on those which are associated with acoustic impedance variations, surface irregularities and to some extent with topography.

None of these types of sonic images contain, however, *range* information, but if the scanning aperture has the capability to measure the time that the information carrying sonic energy package has taken to travel from the target to the aperture, then an 'artificial' sonic image is created in which the range information is replacing the altitude information of the C-mode image. To distinguish it from C-mode it is called B-mode image.

Seeing by sound does not mean, however, only those images which are more or less a faithful replica of the target, but it also means those information patterns which represent the information carried by the sound beam in an unconventional way to which we are not accustomed to in handling optical information. This is called A-mode which differs from B-mode so far that it displays information on range and acoustic impedance properties along a given line of wave propagation, instead of range and azimuth. This type of sound information representation is also called time-amplitude display.

References

[34]FREDERICK, J. R. (1965) *Ultrasonic Engineering* John Wiley & Sons, New York, London, Sidney.
[35]STEPHENS, R. W. B., BATE, A. E. (1966) *Acoustics and Vibrational Physics* Edward Arnold Publishers Ltd., London.
[36]HUETER, T. F., BOLT, R. H. (1955) *Sonics* John Wiley & Sons, New York.
[37]MATAUSCHEK, J. (1961) *Einführung in die Ultraschalltechnik* VEB Verlag Technik, Berlin.
[38]GREGUSS, P. (1968) *Some Design Aspects of*

Ultrasonic Equipment Central Mechanical Engineering Research Institute Report No. M6, Durgapur.

[39]GAMO, H. (1964) *Progress in Optics* Vol 2, North Holland Publishing Company.

[40]GABOR, D. (1965) CBS Laboratories *Seminar on Imaging with Coherent Light,* Stamford.

[41]GABOR, D. (1949) *Proc. Roy. Soc.* **A197** 454–487.

[42]GREGUSS, P. (1966) *Science Journal* **2** 83.

[43]BIBERMAN, L. M. (1973) *Perception of Displayed Information* Plenum Press, New York, London.

41

4   SONIC IMAGE FORMATION

Sonic image sensors are based either on sound-material interaction which results in a proportional change of optical properties of the material or on the interaction, resulting from the image formation, of the acoustic field with the recording material. This then generates an electrical signal proportional to the acoustic intensity at a given point, and their electric properties are used to display the sonic image in a visible form. The advantage of the image sensor belonging to the first group is that they are in general area sensors, i.e. they yield a two-dimensional optical replica of the sonic image directly, while those belonging to the second group have to use sampling, i.e. scanning technique. Since area sensors are generally either less sensitive or have not as good signal-to-noise ratio as scanning sensors, most of the present sonic imaging systems are based on sensors belonging to the second group. Another reason for this is that optical displays with electrical signal input, e.g. cathode ray tubes, are readily available.

4.1   Nonsampled sound images

Area detectors can be divided into two main groups, passive and active ones. In the first case, the information-carrying sound waves are acting on the material of the sensor similarly to light waves when acting on light sensitive materials. In set-ups belonging to the second group, the information-carrying sound waves are merely traveling through the material of the sensor and the mostly reversible changes in the media due to this propagation are read out by optical means, such as reflection, refraction, and deflection of noncoherent or coherent light; they operate therefore generally in real time. The common feature of real area sensors is that they have two dimensions at least of about two orders of magnitude larger than the wavelength of the information carrier and that all pixels are parallel processed.

4.1.1   Sonosensitized photographic plates

As already indicated in the Chapter, *Historical Introduction*, the endeavors to use photosensitive materials as acoustic image sensors were unsuccessful, but it was shown in the early 1950s that they can be sensitized for ultrasonic waves. Arkhangelskii[44] has demonstrated that photosensitive sheets can be sonosensitized by exposing them to white light uniformly, i.e. fogging by diffuse white light. When these sheets are insonified in a dilute developer, the enhanced diffusion of developer into the gelatin of the emulsion at regions of high ultrasound intensities as compared to regions of low intensity causes a black and white sound image. The ultrasonic exposure time varies from one tenth of a second to a few minutes depending on ultrasonic intensity, the concentration and the composition of the developer and, further, on the *photographic* parameters of the sonosensitized sheet.

The threshold intensity for most photographic materials is in the order of about 0.2 W/cm^2, but sometimes 0.05 W/cm^2 can also be achieved, although at the expense of exposure time.

The theoretical resolution of such a sonosensitized film is in the same order as the thickness of the emulsion layer, i.e. between 6–12 micrometers, which is much greater than the wave resolution for most ultrasonic frequencies in practice. This resolution can, however, never be achieved due to the nonuniform streaming caused by the sound pressure-related acoustic wind which is a function of intensity. The resulting blurring effect can be seen in Plate 1 showing a standing wave pattern. The effect of

Plate 1 Standing ultrasonic waves recorded on sonosensitized photographic plate.

streaky disturbances can somewhat be reduced by using lower ultrasonic intensities and a special filmholder having sound transparent wall and containing only a small volume of developer.

To encourage the reader to get involved in sonic imaging and because sonosensitizing methods are rather simple, we give a formula which allows sonopictures to be made with overall quality as good as shown in Plate 2.

43

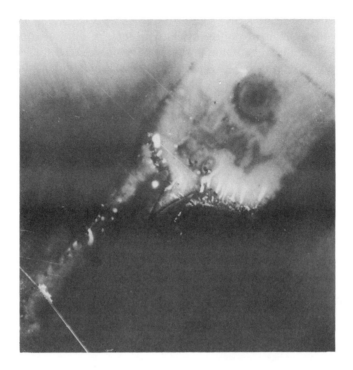

Plate 2 Sound image of a key recorded on sonosensitized plate.

After the photographic film or plate has been fogged by uniformly diffuse white light, it will be exposed to ultrasonic radiation in a solution of dilute sodium thiosulfate and potassium bisulfate. As a result, the silver halide will be dissolved at different points at different rates depending on the intensity of the ultrasonic wave at each point of the plate surface. Since the solution is transparent to visible light, it can easily be judged when the sound exposure is at optimum level. After exposure, the plate is removed and is rinsed with great care in running water, then developed in a developer specified for the plate used, then fixed and dried. Another version of this method is that the prefogged plate is first developed as prescribed by the manufacturer, but only for a fraction of the developing time needed. Then it is washed very carefully. This washed, underdeveloped plate is then used as sonosensitized plate in the solution described above.

It is well known that the human visual system can make many times more distinctions in color than in gray levels, and therefore the question arises whether color films can or cannot be sonosensitized in somewhat similar way to black and white ones. The answer is yes, especially those color films which use two-stage developing technique[45]. It has to be emphasized that in this case color means amplitude-bound information, and not frequency-bound information, as in color photography.

44

4.1.2 Sonosensitive plates based on . . .

Sonosensitive plates are the counterparts of photographic plates, i.e. they are based on different sonochemical reactions. Weissler[46], Greguss[47, 48], have published reviews of sonochemistry, containing several hundreds of references. Although some of these sonochemical reactions have been used to detect and record ultrasonic field intensity distributions, unfortunately no real systematic studies have been made as yet on how they could be utilized in developing sonosensitive plates comparable to photographic plates in resolution, sensitivity and exposure time. Though such an endeavor is an unlikely area for a get rich quick program, there is a chance for potential return and, therefore, the second aim of this book is to arouse the interest in this direction of research.

4.1.2.1 Sonochemical reactions   The first sonochemical reaction which was used to record and see an ultrasonic image was most probably the action of ultrasound on potassium iodide solution manifesting itself in the production of free iodine, which becomes colored if starch is added[49]. The intensity of the purple color depends on the magnitude of the ultrasonic intensity and upon exposure time. Sonosensitive plates based on this reaction can be prepared from normal photographic emulsion. The photographic plate is fixed without development, hardened and washed, and before drying soaked in starch solution for about 24 hours. The concentration of the aqueous solution of potassium iodide in which the sound exposure takes place should be about 6×10^{-5} g/cm^3, and rather good sonic images can be obtained with intensities of about 0.5–1 W/cm^2 and exposure time of several minutes. The resolution is rather good, but the contrast sensitivity is somewhat poor. It can be improved by adding Tinctura saponaria to the film to increase its capillary activity[50].

An improvement in contrast and gray scale can be achieved if the sonosensitizing method described in 4.1.1 is combined with the iodine method[51]. The effect of sound exposure on the emulsion is to render the emulsion resistant to fixing to an extent proportional to the exposure. If the emulsion faces away from the ultrasonic source, the evolution of the sonic image can be observed, because the emulsion turns dark yellow. The quality of the resulting sonic image can then be further improved by fixing the plate after sound exposure in order to clear the insonified part of the emulsion. Then the areas of greater exposure remain essentially yellow, while in areas of intermediate exposure the emulsion is partially clear. There is, therefore, some gray scale in the image. A typical sonic image recorded by this technique is shown in Plate 3.

Studies of the effect of ultrasonic waves on dyes lead to the discovery of the so-called sonochrome substances, i.e. substances whose color changes on ultrasonic irradiation. According to their chemical composition, different factors of the ultrasonic field may act on the compound, but in most cases the oxidizing or reduction effect of the ultrasonic energy is responsible for the color change. Since a detailed list of the different dyes suitable as sonochromes, and the description of their mechanism would go beyond the scope of this book, we refer to the references already

45

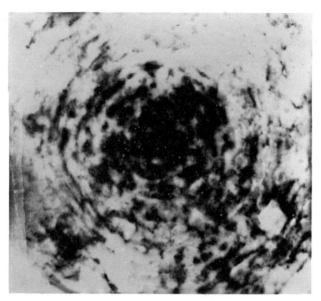

Plate 3   Sound image of an ultrasonic transducer recorded on an iodine-sonosensitized plate.

cited, and to some new ones[52, 53, 54].

The most disturbing property of these sonochrome compounds is not that they are in general yes-or-no indicators, but that they have resisted all attempts until now to soak photographic plates for producing sonosensitive plates. In spite of this fact we mention them here, because they could still be a starting point for developing sonosensitive plates, for instance, the intensity threshold of changing their color may be rather low if instead of the soaking method a new technique is found. Consider also the micro encapsulation technique used to embed liquids or solid particles in compartments varying in size from a few microns in diameter to about 1000 microns[55]. By encapsulating sonochrome compounds they would remain in their liquid state which seems to be essential in most cases to produce color changes.

Microencapsulation provides a further method to produce small compartments which release their contents by pressure or heat to generate conforming reactions leading to image formation of very high resolution. Combining this method with the technique outlined by Ernst and Hoffman[56] may lead to a basically new sonic imaging technique. The method proposed was the 'detonation method' based on the work of Richard and Loomis[57], who found that nitrogen tri-iodide compounds detonate in non-wetting liquids when exposed to ultrasonic radiation. Since Egert[58] produced a light-sensitive printing paper with a very fine dispersion of nitrogen tri-iodide, the decomposition of which is accompanied by a mild explosion if exposed to the flash from a high intensity discharge tube, it may be feasible to develop its acoustic counterpart. The image in both cases is formed by the iodine produced in the micro explosion.

46

4.1.2.2 Sonoelectrochemical Reactions Depending on their intensities sound waves may influence electrochemical reactions, which can be used to record an acoustic image, as demonstrated by the so-called alu-sonophot method[59]. It is based on the observation that in projecting sonic waves on an aluminum sheet which is the anode in an electrolytic cell, the local thickness of the oxide layer developing on the aluminum sheet will be proportional to the sound intensity at the given point. If not too thick, this layer then shows Newtonian interference in oblique illumination.

If the intensity distribution has not been uniform, we shall see the intensity distribution of the acoustic field in colors. Close rings of any one color link points of identical intensity, but separate rings of the same color do not necessarily mean areas of the same sound intensity. The sound image quality depends on the electrolytic condition—the DC voltage applied, the current density, the distance between anode and cathode, and especially, on the composition of the electrolyte. Although exposure time may be short in the order of a few seconds, rather high intensities in the order of a few W/cm² are needed. Plate 4 is the photographic image of the near field of a barium titanate transducer recorded by this technique.

Plate 4 Sound image of a cracked barium-titanate transducer recorded by alusonophot method.

If some object is placed between the ultrasonic source and the aluminum plate acting as an anode, as shown in Fig. 10, the contours as well as any inner inhomogeneity of the object will show up on the aluminum plate. If the intensity of the sound waves passing through the object is attenuated to varying degrees, oxide films

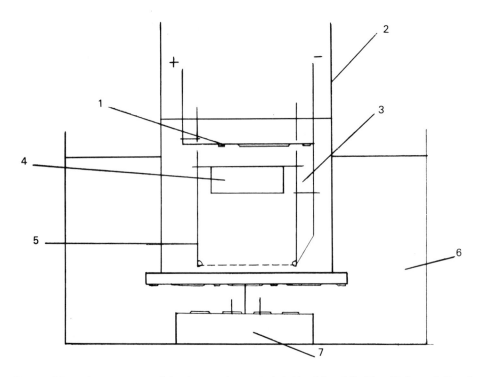

Fig. 10   Schematic arrangement of the alu-sonophot method. 1. Aluminium foil with oxide layer. 2. Vessel. 3. Electrolyte. 4. Target. 5. Frame for acoustic window. 6. Water bath. 7. Transducer.

of various thickness will be formed and the alu-sonophot will have a detailed structure. An L-shaped cork pictured by this method is shown in Plate 5.

4.1.2.3 Sonoluminescence   According to the tradition of designating luminescence by the method of excitation, luminescence which appears when sound waves act on a material has been called sonoluminescence[60]. It can be observed mainly in liquids, but it may occur in gases and solids too. Spengler[61] was the first who suggested the use of this phenomenon to record sonic images. He used 5-amino-2,3-dehydro-1,4, or luminol in alkaline solution. The spectrum of the light emitted was of a broad band with maximum emission at 424 micrometer. The sonic images obtained showed rather good resolutions, interference fringes could be recorded, but the exposure times and the sound intensities needed to make good photographs of the visualized sound image were so high that no practical application of the method could be considered.

Recently, it has been suggested to combine the sonoluminescent detector and a solid-state sandwich type of image intensifier which would allow lower sound intensities and shorter exposure time[62]. A diagram of such a system is shown in Fig. 11. One side of the solution holder is sound transparent, the opposite side, a glass plate,

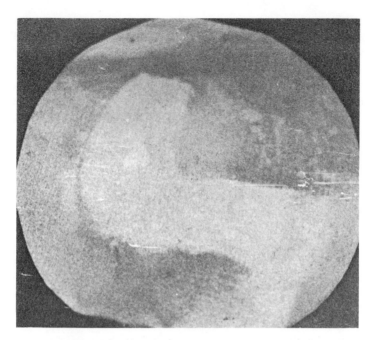

Plate 5   Sound image of an 'L'-shaped cork recorded by alu-sonophot method.

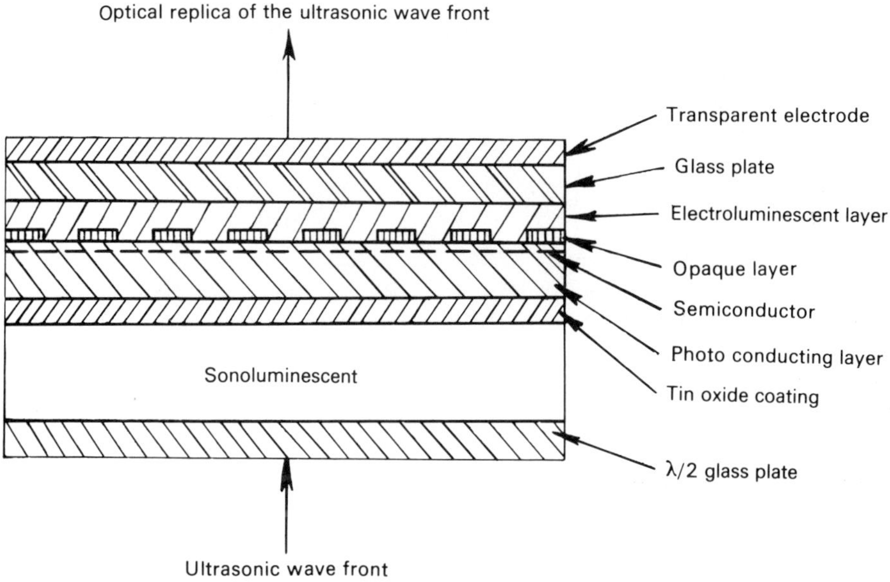

Optical replica of the ultrasonic wave front

Transparent electrode

Glass plate

Electroluminescent layer

Opaque layer

Semiconductor

Photo conducting layer

Tin oxide coating

Sonoluminescent

$\lambda/2$ glass plate

Ultrasonic wave front

Fig. 11   Schematic of the sonoluminescent detector combined with solid state image intensifier.

49

is coated with tin oxide which serves as an electrode. The photoconductive layer, PC, and the electroluminescent layer, EL, are isolated by an opaque film, which eliminates the optical feedback from the EL to the PC film. The second electrode of transparent Au film is deposited on the EL layer. An alternating voltage connected to these electrodes is divided between the PC and EL layers. When ultrasound is not acting on the solution, i.e. there is no sonoluminescence, it has several times higher impedance than that of the EL layer, the voltage therefore on the EL layer is low, and there is practically no light intensity output. When sonoluminescence occurs, the resistance of the PC layer decreases, producing an increased voltage across the EL layer which results in significant light output. Since the sonoluminescent pattern corresponds to the pattern of the sound image, the same electric field pattern is created on the EL layer giving an optical replica of the sonic image. Plate 6 is a photograph of a sound absorbing cross taken by this method with the device shown in Plate 7.

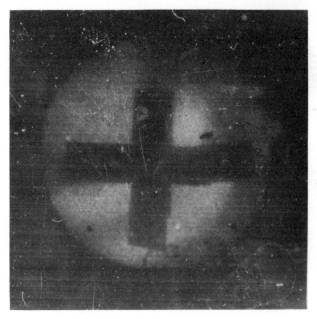

Plate 6 Sound image of a sound absorbing cross recorded with a sandwich type detector of Fig. 11.

Luminescence also occurs if two compounds react, and the resulting light intensity could be modulated by acoustic radiation, according to Mailer *et al*[63]. They suggest construction of an area detector consisting of three sheets, two of them would contain chemiluminescent compounds separated by a relatively fragile material, while the third would be a photographic film. The chemicals should be encapsulated on the sheets, and the complete package could be run between rollers immediately before exposure to ultrasound to break the encapsulation membranes. The chemilumines-

50

Plate 7 The experimental layout of Fig. 11.

cent effect would then take place upon contact of the chemicals, and be modulated by the incident acoustic field. An acoustic image projected on the area detector would cause a local increase in the luminous intensity in proportion to the acoustic intensity incident on the detector. This would be recorded on the photographic film by the contact print process. Processing of the photographic film would then yield the optical replica of the acoustic image.

Not only the intensity of a chemiluminescent reaction and hence the intensity of the emitted light can be modulated by ultrasonic waves, but also the intensity of electroluminescent emission, provided that the electroluminescent layer has a filling material sensitive to ultrasound. This is, for instance, the case if TiO_2 or $BaTiO_3$ or

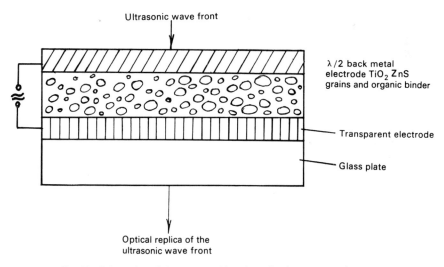

Fig. 12   Schematics of the sonosensitive electroluminescent panel.

51

another, similar, powdered filling material of high dielectric constant is mixed into its matrix layer containing the luminescent powder, that is, into the dielectric medium[64]. The scheme of such a panel constructed by Weiszburg and Greguss in 1959[65] is shown in Fig. 12. This panel, when stimulated with about 220 V, emitted green or blue light depending on the excitation frequency. If the light emission of the electroluminescent panel was just above visibility threshold, and an ultrasonic field was projected on it, depending on local sound intensity the blue or green color of the panel turned into yellow. Plate 8 shows the photograph of an ultrasonic field intensity distribution as seen by the electroluminescent panel. The resolution of the

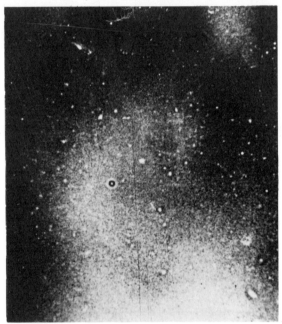

Plate 8 Photograph of an ultrasonic field intensity distribution as seen by the electroluminescent panel of Fig. 12.

panel is rather good since standing waves of 1 MHz frequency can be recorded, only the required intensity is rather high, it is in the order of a few W/cm^2. The contrast in the pictures seems to be poor, but this is a black and white print, while in the original, the yellow sound pattern is on a blue or green background, providing better perception of ultrasonic intensity levels for the human observer than a gray scale.

The mechanism responsible for the observed color change is not yet totally cleared. It is believed that by lateration of the field strength distribution within the dielectric medium the ultrasonic radiation may bring the embedded filling material into a more favorable situation, especially BaTiO$_3$ powder which emits in the yellow-red spectral range. Unfortunately, these experiments have not been continued. However, it would be worthwhile to pursue this idea, and the application of thin-film technique could lead to a new, versatile, acoustic area detector.

4.1.2.4 Thermal effect   It was discussed in 3.5 that as acoustic energy propagates a portion of it may be converted to heat and that the amount of heat generated is a function of sound intensity, i.e. the intensity distribution in the acoustic field is transformed into a corresponding spatial temperature variation over the absorber. If such an absorber is coated with a layer of a compound which changes its optical appearance with temperature, the sound image can be recorded in a visible form. So, if a suitable absorber is coated with a double silver mercury iodide Ag_2HgI_4, the original yellow color of the layer changes to orange, red, or black, according to the intensity of the absorbed ultrasonic wave. Both ultrasonic intensities and exposure times needed to record a good sonic image are in the range where practical application of this method can be thought of: they are below 1 W/cm^2 and 1 second, respectively.

In spite of these good characteristics, this path was not further pursued, probably because the color differences correspond only to large intensity differences, i.e. amplitude resolution is poor. However, as it is indicated in 2.2, multilevel recording may offer little practical advantage in most areas over binary recording and so using this idea a simple and versatile sonic image area detector could be developed without too much effort.

About 9 years ago it was suggested to use cholesteric liquid crystals to map the spatial heat pattern that issues from the absorbed intensity pattern of the sound field[66].

Cholesteric liquid crystals in their liquid phase are colorless, but on cooling to the liquid crystalline phase, or mesophase, they go through a range of colors normally appearing first violet, then blue, green, yellow, red, and finally, colorless again. The rate of change of color with temperature change varies from compound to compound. For a given compound, each color corresponds to an exact temperature. Hundreds of cholesteric materials, both pure and mixed ones are known, several of them have already been tried as acoustic area detectors. To our best knowledge the first experimental work in this field was made by Cook and Werchan[67, 68].

Black polyethylene membrane is generally used as absorbing material, which forms an interface between the sound coupling water and the air. The side of the plastic which faces the air is coated with the chosen cholesteric liquid crystal mixture. Although this seems to be a simple technique, it has such shortcomings which theoretically restrict the application of cholesteric liquid crystals to practical acoustic imaging. This has to be emphasized, since attempts are made over and over to use them for this purpose.

The first problem which restricts here the application of cholesteric liquid crystals is that the temperature rise generated by the absorbed acoustic radiation in the substrate depends, at a given frequency, upon the thickness of the substrate. The substrate can, however, support only a thermal fringe period larger than its thickness. So using a substrate of 0.25 mm thickness, the theoretical resolution would be 4 lines/ mm, while the wavelength of sound at 1 MHz in soft tissue is about 1.5 mm.

This restricting limitation could be improved somewhat by micro encapsulating the liquid crystals in coating materials with high sound-absorbing properties, and using an acoustic impedance close to that of the material on which the capsules are coated. Micro encapsulation should tend to protect the liquid crystals from atmospheric contamination, another limiting factor of liquid crystal area detectors.

To achieve higher resolution the thickness of the material has to be decreased, which technically could be done, but then the intensity of the sound beam—several W/cm^2—would have to be increased considerably.

Another way to improve the sensitivity of this method would be to regulate the ambient temperature to 10^{-3} degrees. As pointed out by Gabor[69], in this case the experimenter might have to wait a month before the temperature equalizes to the point at which he could start to see the sound image.

4.2   Area sensors based on acoustical-optical interaction

As seen in the previous chapters, area detectors based on sound induced physico-chemical effects are not only sensors but yield in general an optical replica of the detected sound field too. Nevertheless, distinction between their functions as sonic image sensors and acoustical-to-optical converters has to be made if we wish to improve their performance. Similar problems are faced in discussing the performance of those area detectors that are based on a direct interaction between a propagating light wave and the acoustic field. Problems arising from this are, however, similar to those which are typical for scanning sensors and will, therefore, be discussed later.

4.2.1   Sound images via schlieren

The German term 'schlieren' by which the sound field recording and visualizing method is generally described means 'streaks'. It refers to the light 'streaks' due to the deflection of a light beam by changes produced in the refractive index of the test medium being studied by the sound waves. The phenomenon was first described and

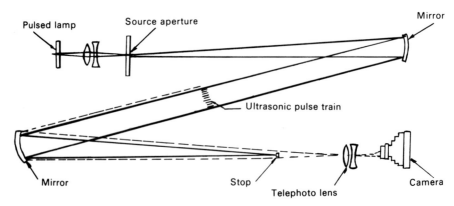

Fig. 13   Basic layout of a schlieren system for visualization of sound fields.

used to visualize sound fields by Toeppler in 1864, and since then it has several times been applied to solve sound imaging problems in science and technology[70, 71, 72]. The principles of this method are easily understood. Fig. 13 shows the basic layout of such a system. A source aperture, illuminated by a lamp, lies at the focus of a concave mirror (or converging lens) so producing parallel light through the working section. This light is collected by a second mirror (or lens) and a stop at the focus of this mirror intercepts all undeflected light. However, light deflected by a sound wave in the working section between the two mirrors (lenses) misses the stop, and a lens behind the stop images this light in an eye piece or camera so giving bright sonic image on a dark background. If a set of color stripes is used instead of the stop, a color sonic image can be recorded in which the color is a function of sound intensity. The sensitivity of the method is very high, it is in the order of 10^{-4} W/cm^2 or less, but this high sensitivity is also its greatest disadvantage.

The system has to be maintained in stable air conditions and has to be extremely rigid. The specimen must be very accurately set-up and aligned precisely to the light beam. In spite of these problems schlieren technique has been successfully used even recently to visualize, for instance, pulsed ultrasonic fields in tube sections[73] for flaw inspection. Nevertheless, the schlieren technique is not recommended for most sound imaging problems to be solved in technology or diagnostics. There is, however, an exception which will be discussed in detail later.

4.2.2 Sound images via photoelasticity

As long as schlieren techniques are based on changes introduced by sound waves in the refractive index of the media, the so-called photoelastic method makes use of the fact that a transparent solid when subjected to a stress other than hydrostatic pressure becomes optically birefringent, i.e. it exhibits a refractive index whose value varies according to the plane of polarization of incident light. Thus, when a sound

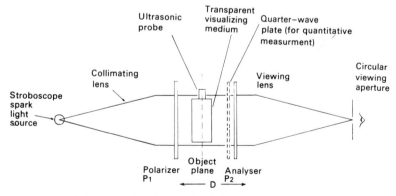

Fig. 14 The optical layout of a photoelastic sound imaging system.

field is projected onto the solid, the spatial sound field intensity distribution will be transformed into a corresponding spatial refractive index distribution. They can be rendered bright against a dark background by placing the solid between and viewing through crossed polarizers. Although its sensitivity is one or two orders of magnitude less than that of the schlieren method, it is worthwhile to consider its application to sonic imaging problems because of its relative simplicity.

The optical layout of a photoelastic sound imaging system is shown in Fig. 14. The light source is positioned at the focal point of a collimating lens. The collimated light beam illuminates the transparent block into which the sonic image is projected. This block sits between linear optical polarizers P_1 and P_2, each of which is held in a rotatable mount. A second lens produces an image of the light source in its back focal plane where the observer positions his eye with the aid of a viewing aperture. The distance between the viewing lens and the plane of the projected sound image to be viewed have to be adjusted to suit the observer's eye.

With the availability of a laser, a polarized and high intensity light source, this method may play an important role in the future of acoustic imaging and, therefore, the reader may find detailed information for derivation of expressions relating image brightness and ultrasonic intensity in the References[74, 75, 76]. Plate 9 shows a sonic image recorded by this technique, using a laser as a light source.

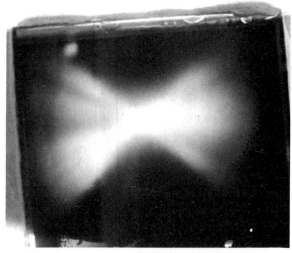

Plate 9 A sonic image recorded by photoelastic method using a laser as light source.

4.2.3 Sound images via nematic compounds

There are compounds, nematic liquid crystals for instance, which exhibit birefringence without being subjected to stress. Nematic liquid crystals differ structurally from normal isotropic liquids only in a spontaneous orientation of the molecules with their long axes parallel. In the absence of special measures, the preferred direction of the long axis is not constant over large areas, but varies with position. Homogeneously aligned layers, i.e. layers having a constant preferred direction behave optically the same way as uniaxial crystals exhibiting positive birefringence, i.e., the ordinary ray has the lower refractive index. The optical axis coincides with the preferred direction of the long axis of the molecule. If an external force is acting on a liquid crystal in a homogeneously aligned layer, the liquid crystal will be disturbed from its equilibrium, but on removal of the force it will return to its previous orientation. The change in free energy which is a function of the applied force and is resulting in deformation can be calculated from the electric constant by considering a deformation which causes a twist in the nematic structure.

Let us suppose that an aligned area of nematic liquid crystals is separated by a distance H from another plane of material oriented at an angle ϕ with respect to the first plane, then a torque will be exerted. According to Zoher[77], its value is

$$T = \frac{a \phi k}{H}$$

where a is the area of the layer, k is a force constant, depending on the type of compound and the method of homogeneous alignment. It is in the order of 10^{-7} dynes. Because of this extremely small force constant the orientation of the liquid crystal can be affected by a small force, for instance, a twist of $90°$ in a 0.1 mm nematic crystal layer requires 10^{-6} ergs/cm^2. This means that a sonic image area detector would *theoretically* respond already to an input of about 10^{-14} W/cm^2.

Since any deformation imposed on a liquid crystal in a homogeneously aligned layer will result in the local change of birefringence, optical rotation, the optical replica of the sound image projected on such a liquid crystal layer would show up.

This idea was tested by Mailer *et al*[78], using a 0.015 in thin nematic liquid crystal layer sandwiched between two glass plates. Since the orientation of the liquid crystals was the result of attractive forces between the glass plates and the liquid crystals, the sensitivity of the system to ultrasonic radiation was rather limited. Furthermore, at higher intensities, a turbulent motion of the liquid crystals produced dynamic light scattering, which introduced uncontrollable patterns in the optical image of the projected acoustic field, reducing resolution considerably. As a matter of fact, the sound-induced dynamic scattering was also proposed for acoustic imaging[79] but was never realized for practical applications.

The acoustical-to-optical conversion cell AOCC[80, 81] probably eliminates the problems mentioned above. It is based on the assumption that if the liquid crystals would float in their free state with their principal axes of all molecules aligned in one direction, i.e. the molecules could be aligned on an axis but not having fixed alignment in other axes, the sound sensitivity of such an area detector could be improved.

This can be achieved if the inner surfaces of the glass plates are coated with a material acting as a 'lubricant', i.e. being a surfactant such as a long-chain fatty acid

Plate 10 Photograph of a standing wave pattern of an ultrasonic transducer in water, as seen on an AOCC area detector.

amine containing multiple amino groups or a quaternary amine[82].

The sensitivity of such an AOCC area detector is at present several milliwatts per square centimeter, but it could probably be improved by at least one order of magnitude or more if more research were conducted in finding new nematic compounds and aligning techniques.

The resolution of the AOCC area detectors reaches practically the theoretical limit, as demonstrated by Plate 10, which is a photograph of a standing wave pattern of a 3 MHz transducer in water. The distance between two dark lines, nodal lines, is about 2.5×10^{-2} cm. Plate 11 shows the interference pattern of two crossing ultrasonic wave fronts.

The gray scale capability of the AOCC, i.e. its dynamic range, is not very large, but this is compensated by the fact that, when it is illuminated with polarized white

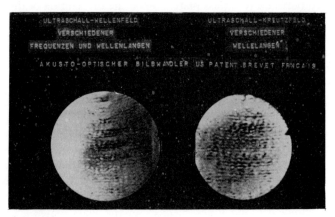

Plate 11 Photograph of an ultrasonic interference pattern as seen on an AOCC area detector.

light, this dynamic range shows up in color, to which the eye is far more sensitive than to gray scale.

The really disturbing disadvantage of the AOCC type of area detectors is that their working mechanism is not yet really understood and, therefore, good area detectors are still made on a more or less trial-and-error basis.

Since we believe that sonic image area detectors based on the strange properties of liquid crystals may have a future if the adequate compound and aligning technique is found, and in order to facilitate the 'where-to-look' procedure we suggest investigation of the following, less known, mostly electro-optical phenomena of liquid crystals.

4.2.3.1 Guest–host interaction In this case, a dichroid dye molecule whose optical absorption of polarized light depends on the orientation of the dye molecule, is introduced as a 'guest' into the crystalline order of the 'host' nematic liquid crystal. The orientation of the dye molecules and their optical absorption are controlled by applying an aligning electric field. Since it has been demonstrated[83] that the optical properties of certain dyes can be influenced by ultrasonic radiation, this type of electro-optical phenomenon of liquid crystals could perhaps be better utilized to develop a liquid crystal acoustic area detector than the voltage controlled optical activity alone which has already been proposed[84].

4.2.3.2 Vertically aligned phase deformation This phenomenon in liquid crystal systems, which shows up in the light transmittance properties of the system, is a function of birefringence of specially prepared liquid crystal cells. Using only 7–10 volts, contrast ratios of 1:1000 can be achieved. Perhaps the orientation of this liquid crystal system can also be influenced by ultrasonic radiation, and so an ultrasonic image projected on such a cell could be recorded.

4.2.3.3 Fast turn-off effect This phenomenon is really a subclass of dynamic scattering, except for that the decay of turn-off times are much shorter. The frequency of electrical excitation for a given liquid crystal type is rather critical: it lies usually in the 600 Hz range. This is, however, the frequency range of the repetition rate used in ultrasonic impulse techniques, so that if the dynamic scattering of these compounds can be influenced by ultrasonic radiation, the effect could be used for recording acoustic images.

4.2.4 Sound images via particle orientation

Sound radiation pressure is a function of sound intensity by which the sound wave is acting on an obstacle, as already discussed in 3.3. If this obstacle is a suspended disc in liquid, it will orient itself to present its minimum surface area normal to the direction of sound propagation. Its real spatial position will be influenced by the viscoelastic forces typical of the liquid in which it is suspended. Since in the absence of an acoustic field the suspended particles are randomly oriented and subjected to Brownian motion, when they are illuminated with a light beam, they reflect a uniform distribution of light, showing a gray background. This is the idea behind the so-called Pohlman cell as acoustic image area detector[85].

As shown in Fig. 15, such an image detector takes the form of a narrow cell filled

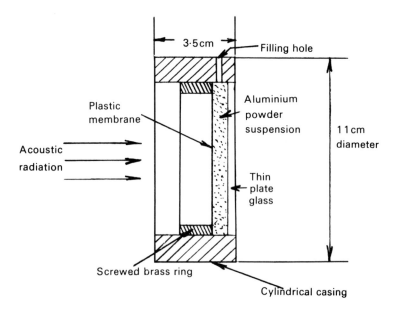

Fig. 15   Schematics of the Pohlman cell.

with a fluid, for instance xylene, in which small aluminum flakes are suspended, one face of the cell is optically transparent, and the opposite face is acoustically transparent. Without insonification the particles are randomly oriented, and a light beam will also be randomly reflected so that the observer will see a uniform matte surface. When an ultrasonic image is projected into the cell, the aluminum flakes will orient themselves according to the intensity distribution in the sound image. It is clear that in this case the observer will see a distribution of light intensity over the area of the cell which is the optical replica of the projected sound image.

This technique, known since 1939, had several comebacks[86, 87] but could never really gain ground. One reason for this is the particle settling under the force of gravity, the other is that its dynamic range is rather limited, about 20 dB. Some improvement can be achieved if a small, typically 20–30 V AC is put across the cell, but even then the contrast range 1:1000 predicted by Fintelmann[88] cannot be achieved. Nevertheless, it would be worthwhile to continue research in this direction too, since the resolution of such a sonic image area detector is limited only by the size of the aluminum flakes. If the flake radius is less than $\lambda/(2\pi)$ the resolution would be limited by wave-optical considerations. Its threshold sensitivity is in the order of a few mW/cm^2 so that it can be considered also for biomedical applications.

4.2.5 Sound images via liquid surface levitation

At present perhaps the most known and most discussed acoustic image area detectors are those based on the liquid surface levitation method, first described by Sokolov in the late 1920s. Very likely the reason for this is that practically everybody who gets involved in acoustic holography starts with this technique. It is based on the observation that a liquid surface will be deformed according to the intensity distribution in the acoustic field acting on it, and that deformation can be visualized by reflecting light from the surface. Plate 12 is such a record obtained by the author

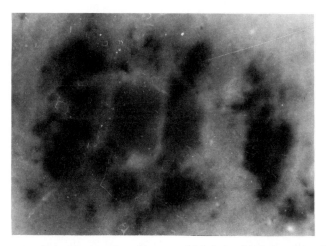

Plate 12 A sonogram obtained by the author with a copy of Sokolov's original arrangement shown in Fig. 1.

in 1950 with a copy of Sokolov's original arrangement shown in Fig. 1. A great number of analyses of this technique have appeared in the literature[89, 90, 91]. Another liquid surface levitation method which is usually relegated to the class of acoustic curiosities, but if modernized could be used for solving several acoustic imaging problems in research and development, is discussed here.

The original Sokolov-technique, which was almost simultaneously re-discovered by Mueller *et al*, and Brenden in 1967, although very attractive, e.g. its threshold sensitivity is in the order 10^{-4} W/cm^2, needs a very elaborate technique to eliminate those unwanted vibrations to which the liquid surface is very sensitive, and, consequently, is rather expensive. In addition, the liquid itself, because of gravitational effects and surface tension, acts as a bandpass filter, typically functioning only in the range of 1–20 cycles/cm so that the information of lower spatial frequency is not adequately reproduced. Further, it is very restricted in aperture, which then also influences its image resolution capability.

The technique we wish to discuss originates from Spengler[92] and can be understood from Fig. 16. F is a disc-shaped reflecting surface having a white central portion

61

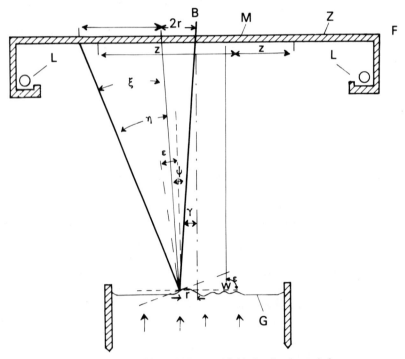

Fig. 16   Schematic of Spengler's sound field visualization technique.

M, with radius r, and an adjacent gray scale portion Z, the rim of the disc being black. This disc is illuminated with a ring shaped fluorescent lamp, having a larger diameter than the edge of the gray scale zone. Placing this set-up at distance d parallel to the liquid surface G, and looking through the aperture B in the center of the disc, the white central portion M shows up. If now a sound field is projected onto the surface of the liquid, this will be deformed. Depending on the deformation at a given point, which is a function of the sound pressure at that point, grades of the gray shade will show up, i.e. the optical replica of the sound field intensity distribution of the sonic image becomes visible.

The threshold sensitivity of this system can be calculated as follows. Suppose that the angle of deformation ϵ, and angle ψ at which the surface is hit from the screen by the reflected light is such that deflection angle μ can be expressed by

$$\mu = 2\epsilon + r/d \qquad\qquad \textbf{35}$$

A particular gray scale point can only be seen through the pupil if ϵ exceeds a threshold value ϵ_g, i.e.

$$\epsilon > \epsilon_g = \frac{Z - 2r}{2d} \qquad\qquad \textbf{36}$$

On the other hand, the height of the deflection, Δh, is a function of ϵ

$$\Delta h = \beta \lambda \epsilon \qquad\qquad\qquad 37$$

where λ is the wavelength of the impinging sound, and $\beta \leq 1$. Further, it is also a function of the intensity difference ΔI at two adjacent points.

Sensitivity S is defined as the reciprocal of ΔI, then

$$S = \frac{1}{\Delta I} = \frac{k}{\beta \lambda \, e_g} = \frac{k \, 2d}{\beta \lambda \, (Z - 2r)} \qquad\qquad 38$$

where k is a constant having the value 7.65[92].

As a consequence of Equation **38** the threshold sensitivity is higher at the rim of the detector than in its center, and this strange feature decreases with increasing screen–surface distance, although the overall sensitivity increases. This may be advantageous in some cases when scattered ultrasonic fields have to be imaged since the intensity of the scattered waves drops rapidly with increasing scattering angle. Nevertheless, threshold intensities in the order of 10^{-2} W/cm^2 can be obtained.

It is a shortcoming of this method that if the deformation amplitude is small as compared to the lateral distance of the two points to be resolved, the sharpness of the edges will then decrease. This imposes limitations on the spatial resolution of the system.

This technique can also be described as a lensless schlieren technique, and as such, it can also yield color sonic images, in which color is a function of intensity (4.2.1). This can be achieved by replacing the gray scale with a set of concentric color rings on the reflecting screen.

Sound-induced deformation of the liquid–gas interface is stationary both in Sokolov's and in Spengler's method. If, however, sound impinges under an angle θ the ripple pattern moves across the surface with a velocity $V = c/\sin \theta$, where c is the sound velocity in the liquid. This sonic image is called dynamic image, and can be regarded as a dynamic hologram as we shall see in 6.5.1. The visualization in both cases is generally different. Static sound images are visualized mostly with a sort of schlieren technique, while dynamic sonic images with scanning technique, therefore they will be discussed in 4.3.1.

4.2.6 Sound images via solid surface deformation

Sound induces deformation not only on liquid–gas interfaces but also on solid–gas or solid–liquid surfaces, such as plates or membranes in contact with gas or liquids. Sonic image area detectors which are based on the formation of ripple pattern on the surface of a plate or membrane, bypass all the problems issuing from the mechanical instability of the interface, poor sensitivity, acoustic streaming effect, and limited spatial frequency response. The scanning approach to the readout of sonic images not only requires complex apparatus to translate the probing beam at high speed, but since pixels of the scanning image are probed sequentially and not parallel, their on-line optical processing is not possible and so an important advantage of the area detectors has to be given up.

The recently developed technique of Fox[93] allows combination of the high resistance to unwanted mechanical vibrations, the high overall sensitivity, the ability to separate the information-carrying signal from statistically scattered light, and high resolution with capability for on-line Fourier processing. The fundamentals of this method are illustrated in Fig. 17. A signal generator SG provides the same signal to

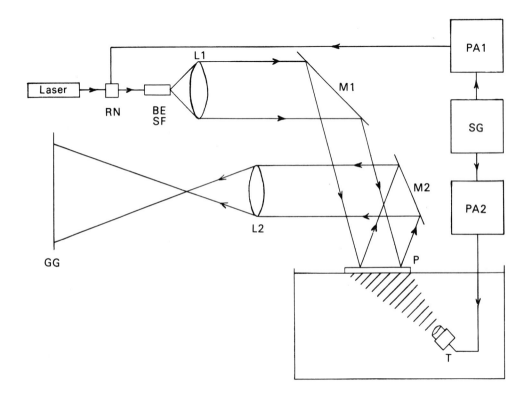

Fig. 17 Arrangement of an area heterodyne optical detection system for sound imaging.

both a Raman-Nath cell RN, and a sound source T. As a result, the laser beam passing through the Raman-Nath cell acquires similar spatio-temporal modulation as the sound source, so that the beam emanating from the cell can be represented by

$$E_\mathrm{a} = E \exp\{i[\omega_\mathrm{l}t + m_\mathrm{m} \sin(\omega_\mathrm{m}t + Kx)]\}$$

where E is the maximum electric field amplitude, ω_l is the radial frequency of the light, m_m is the maximum phase modulation, ω_m is the radial frequency of the phase modulation, K is the acoustic wave number. (Constant phase terms not contributing to the imaging process are dropped.)

This modulated laser beam emerging from the spatial filter and beam expander illuminates the front surface mirror plate on which the sound field is impinging. The reflected wave front is

$$E_\mathrm{c} = [1 + \exp(i\omega_\mathrm{m}t)] \exp\{i[\omega_\mathrm{l}t + m_\mathrm{s}\sin(\omega_\mathrm{s}t + 2\pi f_\mathrm{o}x)]\}$$

where m_s is the modulation index of the acoustic ripple, and f_o is the spatial frequency of the acoustic field pattern. (Constant phase terms not contributing to the imaging process are also dropped here.)

The conversion of the optical phase modulation to amplitude modulation is accomplished by a converging lens L_2 which makes the conventional Fresnel diffraction modulation conversion technique more convenient by introducing magnification and bringing the image plane closer to the detection plane, thus avoiding long optical processing distances.

When a low pass sensor such as a photographic plate is used, then the recorded intensity will be

$$I \simeq 2 + 2m_\mathrm{s}\sin[(\omega_\mathrm{s} - \omega_\mathrm{m})t + 2\pi f_\mathrm{o}x]$$

which, for $\omega_\mathrm{m} = \omega_\mathrm{s}$, is a static representation of the ripple pattern on the plate surface, i.e. it is the optical replica of the acoustic image.

4.2.7 Sound images via Bragg diffraction

In the previous chapters the term 'optical replica of the sound field' has been used, this is correct only if under sound field not sound space is understood, i.e. phase-bound information is not included. Since every area detector discussed recorded only a two-dimensional section of the sound field, so that it would have been more correct to say: these area detectors yielded the optical replica only of a section of a sound field. Bragg-diffraction imaging, however, has a three-dimensional character, thus yielding really the optical replica of the sound field.

The diffraction of light by ultrasound in liquids and transparent solids was predicted already in 1922 by Brillouin[94] but was proven experimentally only ten years later by Debye and Sears[95]. The idea to use this phenomenon for acoustic imaging originated from Korpel[96, 97, 98], and it can be understood from Fig. 18. For the sake of simplicity, only a single point object is shown, radiating a spherical acoustic

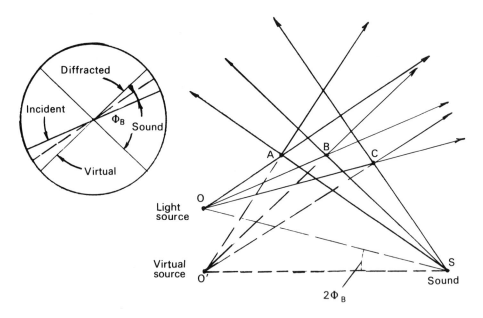

Fig. 18   The basic idea of Bragg diffraction imaging.

wave whose propagation direction is indicated by the arrows that represent the acoustic rays. Light source O emits rays which pass through the sound field, and at points A', B', and C', the light rays make an angle φ with the sound wavefront which are perpendicular to the sound rays. If φ is the Bragg-angle, such that

$$\sin \phi = \frac{\lambda}{2\Lambda} \qquad\qquad 42$$

then the light rays will be diffracted, having satisfied the Bragg condition at the points A, B, and C. The diffracted wave forms a virtual image of the light source at point O. In practice, the light wavelength λ is usually much smaller than sound wavelength Λ, and hence the Bragg angle φ is small. Therefore the light intersects the sound beam in a direction almost perpendicular to the direction of the propagation of the sound, the position of the virtual image O' in practice being close to light source O, and O may be considered as an image of the point source S. If the sound wave is a complicated wavefront coming from an object, the complicated wavefront may be considered to be made up of many spherical wavelets, each having their own virtual source origin. Consequently the diffracted light wave will form many virtual optical images O' producing the optical replica of the acoustic field.

The schematic diagram of such a sound imaging system is shown in Fig. 19. A collimated laser beam is passed through a cylindrical converging lens to form a wedge of light which passes through the water-filled acoustic cell and forms a line focus at

66

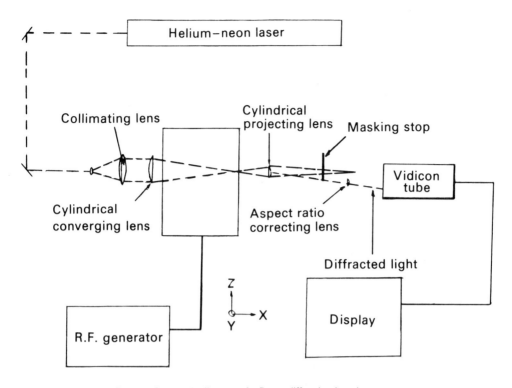

Fig. 19　Schematic diagram of a Bragg diffraction imaging system.

plane P_1 outside the acoustic cell. The sound transducer mounted at one end of the acoustic cell produces longitudinal acoustic waves in the water, which travel through the region illuminated by the converging wedge of light.

There are two possibilities of operation: transmission imaging and reflection imaging. In the first case the object is placed in the region between the transducer and the wedge of light, while in the second the object is placed between the wedge of light and the end of the acoustic cell which is opposite to the transducer. In both cases, the presence of the object results in a complex distribution of scattered acoustic waves causing small fluctuations in the reflective index of the water. These variations of reflective index cause the laser light to be diffracted according to Bragg's law, and produces two image beams on either side of the line focus of the wedge of light (P_1). These beams are the replicas of the scattered sound beam, but unless they are optically processed in the proper fashion, the image derived from them will be distorted. The distortion is due to the different magnifications for the image in the two orthogonal directions. The magnification in one direction is a unity, but the magnification in its orthogonal direction is given by the factor λ/Λ.

67

This distortion can be corrected by projecting the images and the zero order line focus plane P$_2$ with another cylindrical lens. A masking stop is used in plane P$_2$ to block the zero order line and one of the diffracted images. Another cylindrical lens whose axis is rotated 90° with respect to the axis of the other cylindrical lens is placed in the beam to correct the remaining distortion in the aspect-ratio of the image. This image can be viewed directly on a white screen without the use of any other optical image forming device.

This rather simple acoustic imaging arrangement can further be simplified by using a cylindrical acoustic lens in the back focal plane of the object to be imaged and which has to be placed as shown in Fig. 20. In this case, the system does not require additional optical lenses placed behind the sound cell to project the image and to obtain a correct aspect ratio[99].

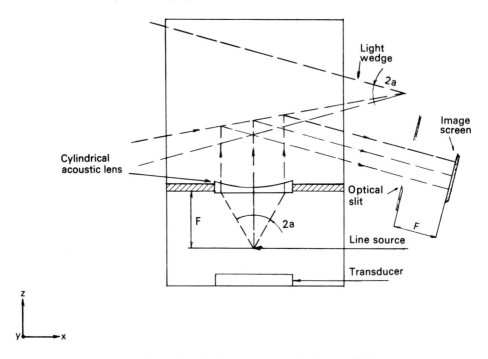

Fig. 20   Bragg diffraction imaging system with an acoustic lens.

The threshold sensitivity of this sound imaging system is very high. It is in the order of 10^{-9} W/cm^2 and can be improved by 5 orders of magnitude, i.e. to 10^{-14} W/cm^2, by using a holographic detection technique[100, 101] which eliminates the effects of the unshifted light acting as a major source of noise in the visualized sound image. The basic idea of this technique is that the frequency of the image forming light is shifted from that of the light originally fed into the system by an amount equal to the acoustic frequency, while most of the background light—which is scattered into the

68

image location—remains at the original frequency and cannot be removed by filtering. However, by making a hologram whose object beam consists of the frequency-shifted, Bragg-diffracted image light, and whose reference beam has a frequency precisely equal to that of the image light, noise reduction can be achieved. The set-up for such a procedure is shown in Fig. 21.

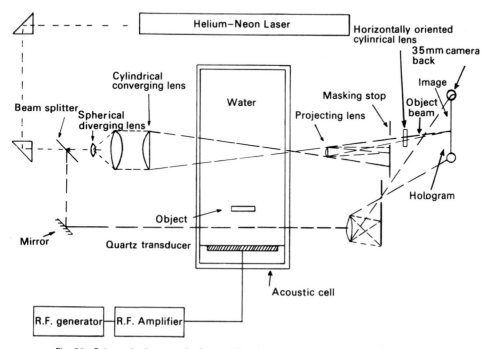

Fig. 21 Schematic diagram of a Bragg diffraction system with holographic recording.

If the image and reference beam are of the same frequency, their interference pattern consists of stable fringes which can be recorded on film. Because of the frequency difference between the noise flare light and the reference light, the fringes produced by interference are not stationary. The individual fringes of this latter pattern will move a distance of one fringe spacing in a period of time equal to $1/\omega$, where ω is the acoustic frequency but at the same time it also represents the difference between the flare light and reference frequencies. If the film exposure time is significantly greater than $1/\omega$, the moving fringes will be averaged out and will not be recorded. When this optical hologram is reconstructed in the usual manner, only a duplicate of the acoustic image is seen, free of the background noise from the flare light.

Although acoustic imaging via Bragg diffraction seems to be a simple technique, it is still in the stage of laboratory development, because—amongst other reasons—

69

its resolution depends upon the orientation of the object structure. It is different for a one-dimensional object oriented so that variation in object structure occurs in direction x, and for another which has an orientation at right angles to this direction. Nevertheless, there is a possibility for the practical utilization of Bragg-diffraction imaging systems in nondestructive testing and medical diagnosis, since pulsed operation is also possible. It could have a future, for instance, in solving inspection problems in nuclear plants[102].

4.2.8 Sound images via Debye-Sears diffraction

A practical disadvantage of Bragg-diffraction imaging is that it can be effectively applied only with sound waves at frequencies higher than a few tens of MHz, but in most technical and medical applications a frequency range of 1–10 MHz is needed, because at higher frequencies the absorption of sound wave is too high. Sato and Ueda[103, 104] suggested therefore to use, instead of Bragg diffraction, the Debye-Sears diffraction of light by ultrasonic waves for sonic imaging at a frequency range of 1–10 MHz. The principles of this idea can be followed in Fig. 22.

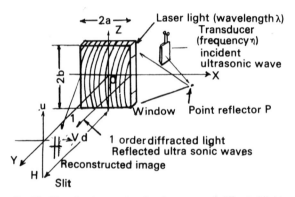

Fig. 22 Wavefront reconstruction by means of diffracted light.

A pair of glass plates is installed in a tank to form a window and the point P to be visualized is insonified by continuous waves of frequency Ω. The ultrasonic waves scattered by the point reflector propagate towards the window and pass through a space between the glass plates. Monochromatic light of wavelength λ is then directed through the window and cross the ultrasonic beam. A part of the incident light is diffracted by the ultrasonic waves and an ultrasonic image of P is formed by the diffracted light at a plane distant d from the window. It is further assumed that the space between the glass plates is so narrow that the variation of sound pressure along direction y is negligible. The image along direction x corresponds to the diffraction pattern of the slit of width 2a, but the central point shifts to $v = x_o$. Consequently, the reconstructed image of the point object is formed at $v = z_o$ and $v = x_o$, however, the point image is lengthened in direction v. In order to shorten the spread of the point image along direction v, the narrow slit is put in the reconstructed image as illustrated in Fig. 22.

A serious drawback of this imaging techinque is, however, that it gives only fine images of objects which are small in direction x, otherwise degradation of the images occurs resulting from superposition of images along direction x. The resolution in the range direction can, however, be improved if linear FM ultrasonic pulses are used for insonification. In this case the reconstructed image of the point reflector becomes a spot centered at $\mu = z_o$ and $v = x_o$, and then the range discrimination power is also greatly improved. Since ultrasonic waves propagate with the velocity of sound, pulsed operation of light is necessary to illuminate the window at the instant when the sound pressure distribution is given by

$$\frac{k \sin(mr^2/c^2)}{r} \qquad\qquad 43$$

where c is the sound velocity, $r = [(x_o - x)^2 + (z_o - z)^2]^{1/2}$ and k is a constant.

With this trick the resolution in direction x is improved to

$$S_x = \Lambda x_o/a \qquad\qquad 44$$

where a is the dimension of the glass plate in direction x. The resolution in direction y will be

$$S_y \geq 2\rho[l + l_2/l_1)] \qquad\qquad 45$$

where 2ρ is the width of the slit, l_1 its length, and l_2 is the distance of the point from the slit. The resolution in direction z is given by

$$S_z > \Lambda x_o/b \qquad\qquad 46$$

where Λ is the wavelength of the ultrasonic waves and b is the dimension of the glass plate in direction z.

When designing a sound imaging device based on Debye-Sears diffraction one has to keep in mind that the sound image of the object is formed just in front of the focal plane, and is so small that another lens has to be used to magnify it. The magnified image is formed on the image plane, and by placing a slit on it a *slice* image of the object corresponding to the image of the object at $y = 0$, $x = x_o$ obtained, i.e. it is one-dimensional. It is therefore necessary to move the object in direction y. Similarly, if a photograph of the image plane in direction y is required, the camera must be moved synchronous to the movement of the object in order to get a sound image corresponding to the cross section parallel to the y–z plane. The self-explanatory schematic of such a sound imaging system is shown in Fig. 23[104].

71

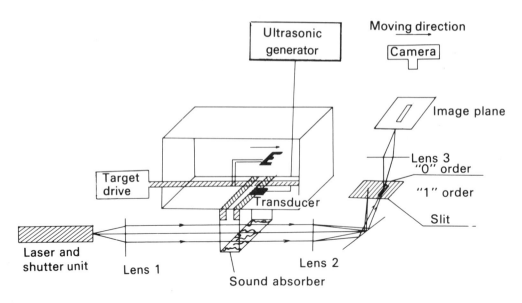

Fig. 23 Schematic diagram of an ultrasonic imaging device using Debye-Sears diffraction of light.

4.3 Sampled sound images via . . .

The most effective way of describing a visualized sampled sound image and to point out the ways in which it differs from a visualized nonsampled sound image is to compare both with the theoretical sound image, the optical replica of which they pretend to be.

Visualized nonsampled sound images—as seen in the previous Chapters—have been formed by causing an intensity pattern on an area sensor to view the entirety of the insonified scene. In the case of sampled sound images the creative act is that of defining a set of resolution elements for the image. The visualized sampled sound image is fully defined when to every element in it has been assigned a two-dimensional light intensity distribution.

How do visualized sampled sound images differ from visualized nonsampled ones? The difference lies not only in the fact that the visualized sampled sound image requires that the sampling aperture, or apertures, be interrogated in a known, precise manner, but also in the strange feature that the sound image to be visualized is generally formed point by point during the scanning procedure itself, i.e. the insonification of the target is a sampling procedure too, in contrast to the visualized nonsampled sound images, where every point of the target has to be insonified simultaneously. This is why visualized sound images may be sampled, but is not itself a sampled sound image, as we shall see in 4.3.1.

The mode of insonification has, however, an important effect upon the sound image itself. Parameters, such as contrast, are difficult to control, and we can never be sure whether the intensity of a given sound image point is a result of the summation of scattered waves or of specular reflection. Using point-by-point insonification, these problems are mostly eliminated, since each sampling element is an independent information channel, however, the functioning of these sampling apertures relates many parameters in a very complex way to the quality of the invisible sound image as well as to the final visualized sound image.

To eliminate possible misunderstandings it is necessary to define certain terms as they will be used in the following chapters. The term 'sampling' refers to sampling in space. There are innumerable ways in which sampling can be done. Fig. 24 shows three of the possible arrangements.

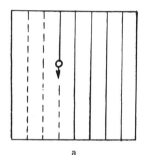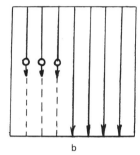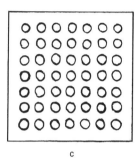

Fig. 24   Three types of detector arrays.

In Fig. 24a, a rectangular raster scan of the scene is made with a single scanning aperture, e.g. piezoelectric crystal. In Fig. 24b, a number of apertures arranged in a single line are swept across the target, and in Fig. 24c, a static two-dimensional array of apertures is indicated. In such a case, if the area of each aperture is identical with the area of the sound image resolution element, scanning occurs only in the visualization process itself.

The use of a single sampling aperture or the use of more than one scanning aperture to collect the set is not fundamental, it merely concerns the extent to which the picture elements, the pixels are processed in parallel. It is important, however, that the image flux is not continuous in the two coordinates, but contains periodic elements. An arrangement limiting aperture positioned to a fixed number of uni-formly spaced points in the image frame is termed a *point raster*. An arrangement providing continuous operation position along uniformly spaced parallel lines is termed a *line raster*. The raster constant n_r specifies the number of aperture positions in the length unit of a geometric arrangement of points or lines. It does not specify, however, the dimensions of the 'points' or 'lines' themselves, which are determined by the geometry of the sampling aperture used with the raster process.

The process of dividing sensed sound intensity into discrete intervals for multilevel, gray-scale representation is referred to as 'quantization'.

To answer the question 'What does sampling do to the invisible sound image quality?' and 'What must the sampling frequency and format be to minimize the deterioration of the invisible sound image?' the following three factors affecting these problems have to be analyzed:

a) the number of samples per sound image,
b) the signal-to-noise ratio per sample,
c) the generation of spurious signals by the sampling process.

The complete evaluation of these factors is beyond the scope of this book, nevertheless, we have to give some guidelines to compare sound imaging techniques. The reader who wants more details should read the Montgomery Research Paper[105] and the Hildebrand and Brenden book[106].

Let us assume that sampling and acoustic signal $I(t)$ every T seconds, we get a set of values $I(mT)$, where $m = \pm 0, 1, 2, \ldots$ The question is what do we know about $I(t)$ on the basis of the sampled values $I(mT)$. Suppose that the samples $I(t)$ are sinusoidal, then, for values of $f \leq \pi/T$, the set of sampled values for the functions $\sin\{2\pi t[f + (m/T)]\}$ is the same for all positive and negative integer values of m. Thus, on the basis of samples spaced T seconds apart, we are unable to distinguish between sinusoidal frequencies $f + (m/T)$ for integer m. The set of frequencies $\{f + m/T\}$ is called the set of aliased frequencies, and may play an important role, especially in acoustic holography (5.4.1). Further, it can be shown that if the signal is sampled, every T seconds and the original signal $V(f)$ is band limited with bandwidth $B = 1/2T$, then $V(f)$ can be exactly reproduced from the sampled values by passing the sampled values through a bandpass filter of bandwidth $1/2T$. At the risk of oversimplification, it can be said that it is necessary to have at least two samples within each cycle of a given frequency in order to detect the frequency as shown in Fig. 25. This is the so-called Nyquist criterion.

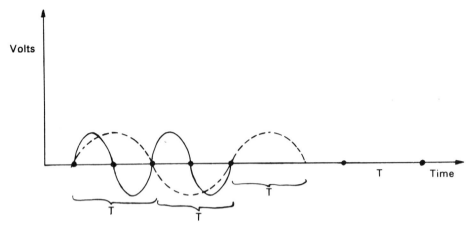

Fig. 25 Sampling relationships.

The reader should not interpret the sampling procedure as a bandpass filtering operation, although there is such a tendency in the literature. It can be shown[107] that higher frequencies are not filtered but added to lower frequencies. Even if the reconstruction function is a bandpass filter with zero response outside the passband $|f| \leq 1/2T$, higher frequencies are aliased or added to the frequencies in the passband. Only if the original signal $V(f)$ is band limited, can we ensure in principle perfect reproduction of $V(f)$ from its sampled values, i.e. can we detect the real sound image that is identical to the nonsampled sound image.

Here we come to a real difference between sampled and nonsampled sound images, which can be further discussed with the aid of Fig. 26. For the sake of simplicity a

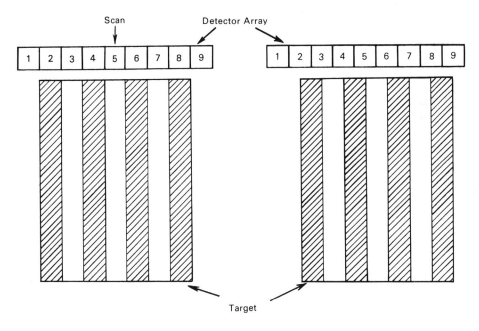

Fig. 26   Effect of scanning spot position.

bar type target is shown, but the effect to be discussed holds for any type of target. According to Fig. 26a a detector array is swept past the target in direction x, then the odd number detectors will show zero signal and the even number ones a maximum signal. The modulation transfer function MTF of the detector array in the vertical direction due to the scan motion unity is at this frequency, and goes to zero only at twice this frequency. In Fig. 26b, the detector array is displaced 1/4 of a target cycle. When the array is scanned in this position, the output of each detector is identical and thus the MTF is zero. Although the array obeys the Nyquist criterion, nevertheless, there is an anomaly in the response if the phasing of the array with respect to

75

the target is unfavorable. This kind of effect is more apparent if we consider the MTF in the two orthogonal directions. This means that sampled sound images tend to be sensitive to target orientation which is usually not the case for nonsampled sound images.

4.3.1 Laser beam sampling

Laser beam scanned sound images are typical examples of visualized sound images which may be sampled but are not themselves sampled sound images, i.e. every point of the target has been insonified simultaneously. This technique can be used to read out sound-induced deformation of a surface which can be either static or dynamic. As already discussed in 4.2.5, sound-induced static deformations are generally read out by some sort of schlieren technique. Although similar methods can be used to read out dynamic images[108], laser beam scanning offers not only the advantage of discriminating between dynamically and statically scattered light (which reduces the available contrast) but because the former is Doppler-shifted by the sound frequency, and because the electrical input is at the sound frequency, electronic image processing in phase and amplitude can be accomplished in real time. Further, using a static method, the target cannot be located close to the surface, which is a serious disad-

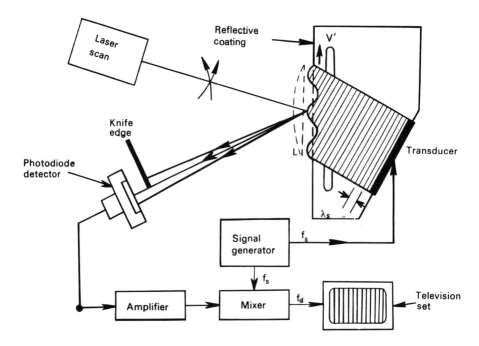

Fig. 27 The principles of operation of a laser scanned system.

vantage especially at higher frequencies where any extra passlength causes excessive attenuation.

The principles of operation of laser scanned systems may be understood by looking at Fig. 27 which is the basic layout of an arrangement developed by Whitman and Korpel[109]. The collimated beam of light emerging from the aperture of the laser beam deflector is focused by lens L_1 on a cover plate having a light reflecting layer on which the sound image is projected. The focal point is scanned over the surface and is reflected to lens L_2 which forms an image of the exit aperture in the plane of the knife edge E. As a laser beam is scanned across the pattern of moving ripples, the amount of light reaching the photocell varies because the position of the aperture image relative to the slit depends upon the slope of the wave surface at the position of the light beam at that instance. The interaction producing the signal can be looked upon as a geometric light beam deflection phenomenon or as a diffraction phenomenon. Displacement amplitudes at the solid surface are of the order of 10^{-8} cm, and therefore require much more sensitive detection methods than those used in the liquid surface technique. This, however, is not a disadvantage because using a focused laser beam and an efficient photocell detector the required sensitivity can easily be achieved and at this cost all problems issuing from the instability of liquid–air interfaces are overcome.

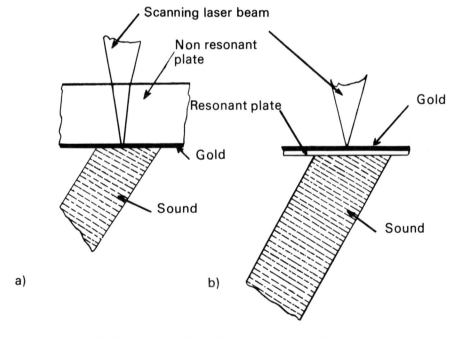

Fig. 28   Two alternatives of laser beam reflecting surfaces.

According to Korpel and Kessler[110], one has to choose practically between two alternatives concerning the design of the laser beam reflecting surface:

a) a thin cover plate resonant at the nominal cycle of incidence, with front surface mirror, or

b) a thick sound-absorbing, transparent, nonresonant cover plate with back surface mirror.

The two possibilities are shown in Fig. 28. The drawback of solution *a* is that at high sound frequencies the plate becomes very fragile and in operation it is limited to the critical angle of total reflection, while in case *b* we may run into the problem of nonuniform insonification of the target, due to the different absorptions between transducer and object. As long as this method is not used at too high frequencies, up to 100 MHz the variation in intensity across the target can be kept about 5 dB which in most cases can be tolerated. However, at higher frequencies this effect rapidly becomes intolerable and a new insonification technique has to be found.

The described laser scanning technique can be applied, for example, to the problem of inspecting structural materials[111, 112] without using a separate cover plate. The surface of the material to be investigated serves as a reflecting surface. In this case, as shown in Fig. 29, the surface wave is excited by a wedge transducer and the light beam is scanned over the vibrating surface of the material by mechanical mirrors in synchrony with an electron beam being scanned over the face of a cathode ray tube.

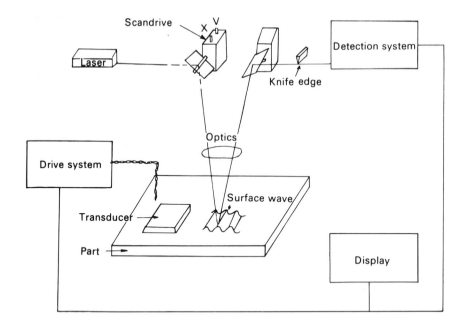

Fig. 29 Visualization of surface waves for nondestructive testing after Allers *et al*[111].

By comparing the phase of the received signal to that of the transmitting transducer in a phase sensitive detector a signal is obtained which varies as $\cos[(2\pi x/\lambda) - \theta]$, where λ is the surface wavelength, x is the position, and θ a constant instrumental phase shift. Stationary displacements of wave crests and troughs are produced by modulating the intensity of an oscilloscope trace by the phase sensitive signal at driving the x and y oscilloscope amplifiers with analog signals proportional to the location of the laser spot. A correcting lens ensures that the reflecting light beam will strike the knife edge and photodiode when they are placed in its focal plane, independent of the scanned angle. Since the surface waves are scattered by the abrupt discontinuities in properties, the visualized sound image will provide a characterization of those physical and chemical properties which are affecting the elastic wave velocity in the region near the surface.

4.3.2 Mechanical sampling

The visualization of a nonsampled sound image by mechanical scanning was first proposed by Sokolov[113]. He suggested to use a piezoelectric plate as area detector, and to explore the voltage distribution replica of the sound image by a spiral arrangement of capacitance probes attached to a Nipkow wheel as illustrated in Fig. 30. When a signal is presented, the gas discharge tube ignites and its light passes through a hole in the Nipkow slutter to illuminate the screen. In this way, a television-like picture of the sound image projected into the piezoelectric plate is produced on the

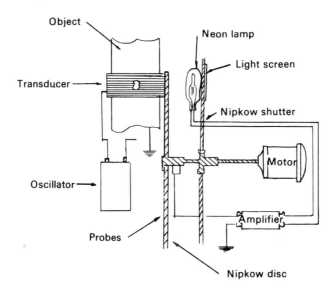

Fig. 30 Schematics of Sokolov's mechanically scanned piezoelectric area detector.

light screen. Although this idea has been changed to scanning the electric replica of the sound image with an electron beam (see 4.3.3), this technique existed for a long time in equipment for seeing by sound underwater, and still exists in some very cheap sonars for fishermen.

4.3.2.1 Linear scanning   Visualization of sound images via mechanical scanning means in most cases a method in which the insonification of the target is also a sampling procedure, but is performed not with continuous sound waves but with sound pulses, generally in reflection mode. There are several scanning configurations producing either B-mode or C-mode sound images.

The simplest scanning configuration is linear scanning, in which the scanning aperture, the transducer is physically moved linearly in direction normal to the beam axis, thus causing the sound beam to scan through a plane section. The result is a B-mode image, which has an excellent resolution in the direction of sound propagation but the resolution perpendicular to the direction of sound propagation, the lateral resolution, is about one order of magnitude lower.

One of the problems with this rather simple sampling method is that if irregular objects have to be insonified, a coupling water bath is needed to ensure the 90° angle between the axis of the insonifying beam and the scanning aperture. The other is that specular reflecting target points are imaged only if they are perpendicular or nearly perpendicular to the insonifying beam. These problems can be partially overcome if the scanning aperture interrogates every point of the target successively from two or more directions. This procedure termed *compound scanning* needs considerably more time than a simple linear scan, and may also lead to degradation of the visualized sound image definition, as a result of any imperfections in the geometrical transfer between target and image space because of the nonaccurate knowledge of sound propagation characteristics in target space.

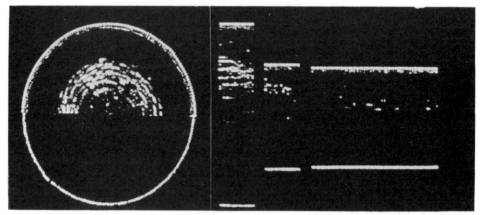

Plate 13   Sound image recorded by mechanical linear scanning for NDT purposes (*courtesy* of Krautkramer Ultrasonics, Inc.)

In spite of the above mentioned problems, mechanical linear scanning techniques can well be used in nondestructive testing (Plate 13) or when static scenes have to be imaged by sound. However, scenes of dynamic nature as in medical diagnostics will yield blurred sound images.

To achieve dynamic visualization, i.e. to supply a real-time sound image without losing the advantage of line scanning another sampling procedure has been proposed[114–117], the concept of which is illustrated in Fig. 31. The scanning

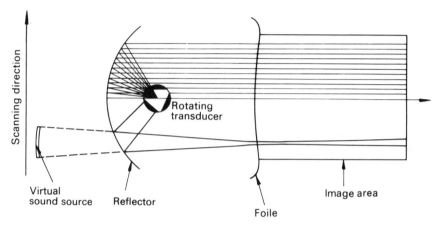

Fig. 31 Schematics of a system using mechanical line scanning to achieve dynamic visualization of the sound image.

aperture, the ultrasonic transducer is placed in the focal point of a parabolic mirror, the emitted sound beam is reflected and projected into the space to be insonified. During the rotation of the transducer the mirror is progressively covered, for instance in 70 ms, so that the section plane, the sound image of which we wish to visualize, is scanned by the ultrasonic beam 15 times per second. If the ultrasound is emitted, let us say, 2000 times per second, the visualized sound image will be made up of 140 lines. With a line spacing of about 1 mm, this leads to a picture width of 140 mm. The speed of operation depends mainly upon the vertical dimension of target space to be imaged. The first requirement is that the interrogating pulse be spaced apart by at least the time taken for echoes to return from the farthest target point of interest. Thus, assuming a sound velocity of 1500 msec^{-1} and a maximum range of 250 mm, the theoretical repetition rate will be 3 kHz, which, however, can never be achieved, due to the delay line of the transducer-mirror-boundary of target space, and vice versa.

Another version of this principle is shown in Plate 14[118]. The rotating transducer is replaced by an oscillating (or rotating) small acoustic mirror in the focal point of the parabolic mirror, and the sound is projected in this oscillating mirror, thus achieving the same linear scanning pattern as with the set-up of Fig. 31. A distinct advantage of this arrangement is the easy changeability of the transducer, i.e. the

Plate 14   Real-time linear scanner after Greguss[118].

variability of frequency of the insonifying beam.

4.3.2.2   Sector scanning   There has been considerable interest over the past decade in the development of rapid sampling methods. Since the oscillation (or rotation) of the transducer is the fastest (and the cheapest) mechanical means to change the position of the interrogating sound beam, a lot of effort has been devoted to reduce the length of the delay line and so to increase the depth of field and the number of frames scanned per second. This goal can be achieved if the linear scanning pattern is given up and a sector scanning pattern is introduced. This approach presents, however, some problems such as

a) frame rates and mechanical oscillations are limited by inertia, balance and motor weight associated with the eccentric drive linkage,

b) fixed sector width, and

Plate 15   Servo-controlled ultrasonic sector scanner[123].

c) concentration of scan lines at either edge of the sector raster owing to the nonlinear scanning velocity.

These problems have been approached in a variety of ways and with varying degrees of success[119–122]. The solution we find to be one of the most advanced is that of Matzuk and Skolnick[123], and is shown in Plate 15. The transducer element oscillates in a small cylindrical case filled with degased oil and having a sound transparent window. The single moving part consists of

a) a lead-metaniobate transducer sealed to an aspheric acoustic lens which simulates the virtual line source principle, achieving large depth of field with high resolution,

b) a permanent magnet pivoted on needle bearings to act as an armature of a rotational torque motor, and

c) a semicircular eddy current position sensing vane structure enabling the transducer angular position to be electronically monitored during the close loop scanning cycle.

An electronically programmed wave form controls the instantaneous position of the oscillating transducer element of low inertia. This wavefront is generally of sawtooth or triangular shape, but may employ an electronic comparator circuit to

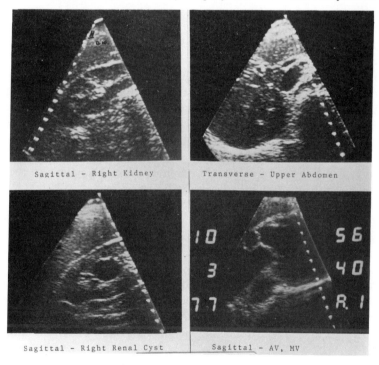

Plate 16 Sound images obtained with high resolution programmable servo-controlled ultrasonic sector scanner after Matzuk and Skolnick[123].

identify a substitute of DC voltage level in lieu of the recurrent wave form. Such a combination of control signals enables the angular position of the transducer to show

    a) variable frame rate from 0–30 frames per second,

    b) variable sector width up till 60°,

    c) uniform line density within the sector raster.

Some visualized sound images obtained with this high resolution programmable servo controlled ultrasonic sector scanner are shown in Plate 16.

It is usually thought that the visualized sound image produced by such an oscillating transducer assembly has to have a sector format. Using the method of Matzuk and Skolnick, however, by time-gating and delaying the scan raster wave form to occupy lower portions of the depth of the field, a nondistorted trapezoidal image format is also possible to use.

Visualized sampled sound images are sampled in reality twice, once during target insonification and, then, when sampled for visualization. From this fact follows that the quality of the visualized sound image depends primarily upon the sampling procedure during insonification, since the quality of the optical display (see Chapter 5) is in general better than the resolution required by the Nyquist criteria for the insonifying beam. As already indicated, B-mode sound images suffer from poor lateral resolution, which can be, however, somewhat improved by using a focused sound beam for insonification or by using suitable sono-optics, but even then the good lateral resolution is limited to a small depth of focus. Therefore, the practicality of this method is rather restricted.

The depth of focus problem can be, however, eliminated by using the insonification method developed by Burckhardt *et al*[124]. How this can be done is illustrated in Fig. 32. The central portion of a spherical transducer is eliminated and the remaining annular part is divided into eight segments. Each element of the ray emits a fan-shaped beam and on the axis the individual waves from all the ring elements add up in phase, and this gives a large amplitude, whereas outside the axis the phases will be different and the resulting amplitude much smaller. All points on the axis have this property that the individual waves emitted from all ring elements add up in phase and, therefore, there is no depth of focus problem.

The disturbing effect of the smaller amplitudes around the axis can further be reduced by emitting with two opposite 90° segments of the ring and by receiving with the remaining two. Since the same annular transducer is used for reception of the beam reflected from the target points, it can easily be incorporated into a smaller sector scanning head as described previously. With this method not only the lateral resolution but also the sampling information from specular reflecting target points are improved, because the annular transducer assembly subtends a considerably larger angle as seen from the reflecting point, than does a conventional one.

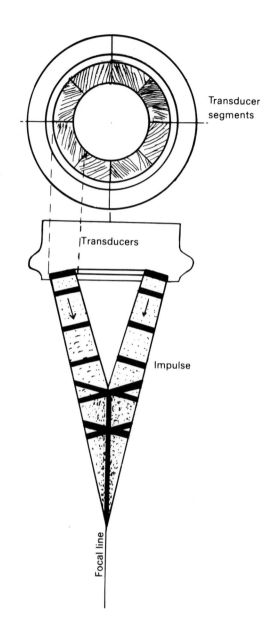

Fig. 32   Annular transducer and the ultrasound beam emitted. Ultrasound is focused into a focal line.

4.3.3 . . . Electron beam sampling

The basic features of the visualization of a sound image by electron beam scanning emanates from Sokolov[7]. This image is a result of a nonsampled sound image projected onto a piezoelectric plate yielding its electrical replica, which is then scanned by an electron beam generating a video signal which represents the sound image. During the last four decades many different attempts have been made to use this idea for seeing with sound in real time[125–132]. Difficulties in the practical application of the proposed converter tubes arose mainly from the fact that the size of the insonified scene is limited by the aperture and by the maximum angle of detection of the face plate. It has to be emphasized, however, that this visual field limitation is determined by the detector aperture and not by, as it is sometimes attributed to, the total reflection angle of sound waves from water (which is usually used as coupling medium) impinging on the face plate[133]. Using mosaic piezoelectric face plate this limitation can be reduced. Plate 17 shows such a mosaic face plate developed by Brown *et al* at the EMI Central Research Laboratories[134], while Fig. 33 illustrates the construction of the tube. The sound image impinging at the mosaic quartz plate generates piezoelectric voltages. The electron gun generates a high

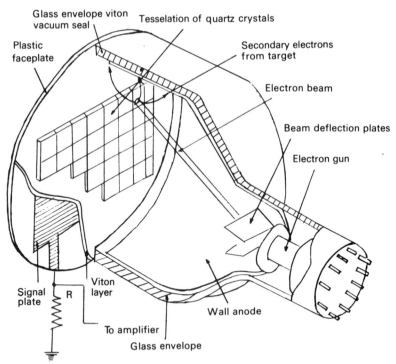

Fig. 33   Face plate construction and electrode layout of the high-resolution sensitive image converter of the EMI Centre Research Laboratories.

86

Plate 17 Photograph of an assembled face plate of the high resolution, sensitive ultrasonic image converter of EMI Central Research Laboratories[134].

velocity electron beam, which is electrostatically focused and scanned over the face plate. It addresses some small part of the quartz plate corresponding to the area of the beam, which defines the size of the visualized picture element, the pixel. Since,

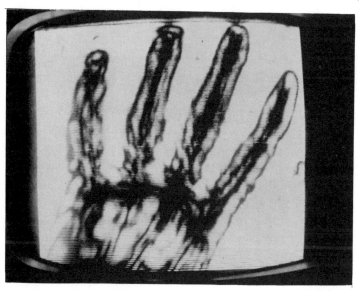

Plate 18 Visualized shadow sound image of a hand placed in the vicinity of the face plate and insonified with 2 MHz, after Brown *et al*[134].

however, the cross-section of the focal point is less than the wavelength of the insonifying beam, it really does not affect the resolution of the sound image, which is limited more by the sono-optical system projecting the sound image onto the face plate. Plate 18 shows the visualized shadow sound image of a hand placed in the vicinity of the face plate and insonified with 2 MHz. Most of the apparent detail seen in the image is due to diffraction of acoustic waves around the bones and flesh contours. A better resolution could be achieved only by using carefully designed corrected sono-optics. In this case, the cone of energy falling on the face plate could extend over an angle of nearly 40° in contrast to 20° without it.

The threshold intensity of the tube is in the order of 10^{-4} W/cm^2 (the theoretical value is 10^{-11} W/cm^2), and one may wonder why it is not used, for instance, in medical diagnostics. The reason for this is that it operates only in transmission mode. The feasibility of the idea to use piezoelectric face plate not only for reception but also for generation of sound waves to build up a visualized sound image was tested by Dubois[132]. In this case, the conversion plate for transceiving must compromise

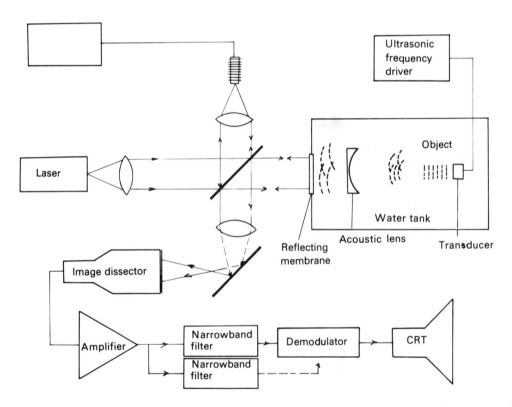

Fig. 34 Schematics of the layout of the arrangement for electron beam scanning of sound-induced light pattern.

between two conflicting requirements. As a transmitter, it should radiate with a narrow beamwidth and high power; a large area of excitation is needed. As a receiver, it must be acoustically and electrostatically unloaded over as large an area as possible. Although Dubois obtained by this technique recognizable sound images, substantial development work is necessary in the design of the face plate and lens system in order to obtain really practically useful images with an acceptable signal-to-noise ratio.

4.3.3.1 Electron beam scanned sound-induced light pattern According to Green[135], to visualize nonsampled sound images, not the electrical field replica but the optical replica created by the sound image on an optically reflecting elastic surface should be scanned by an electron beam. As illustrated in the scheme of Fig. 34 the sound image, projected onto the elastic surface 6, is converted into an optical wave field by a Twyman-Green type interferometer. Laser 1 provides a coherent light source for the interferometer, and the laser beam collimated by lens 2 is split into two orthogonal beams of light by beam splitter 3. The interferometer is arranged so that one of the split beams passing through beam splitter 3 be incident on the outer reflecting surface of membrane 6, on which sound image is projected. The transmitted light beam 4 illuminates the entire surface 6, and is phase modulated by the oscillating perturbations induced at the interface by the acoustic waves to form the object beam. The object beam 4 reflected back from surface 6 is deflected by beam splitter 3 to lens 5. At the same time a portion of the coherent light from the laser source is deflected by beam splitter 3 through a lens to retroreflector 7 to form a reference beam. The reference beam reflected back from retroreflector 7 is transmitted through beam splitter 3 to lens 5. The recombined object and reference beams are imaged by lens 5 onto the face of image dissector 10 to form a time varying interference pattern corresponding to the sound image projected onto surface 6.

The amplitude of the superimposed wave field incident on the image dissector $(x,y$ plane) may be expressed mathematically, using complex notation as

$$U(x,y) = \exp\{i\gamma(x,y)\} + \exp\{ikA(x,y)\sin[\omega t + \theta(x,y)] + i\beta(x,y)\} \qquad \textbf{47}$$

where $k = 2\pi/\lambda, \lambda$ is the wavelength of the light emitted by the laser, $A(x,y)$ and $\theta(x,y)$ are, respectively, the amplitude and phase of the sound image induced perturbation of reflecting surface 6. Further, $\omega = 2\pi f$ where f is the ultrasonic frequency, and $\beta(x,y)$ represents undesired phase variations of the object beam resulting from unequal optical pathlength. Such undesired variations are also presumed to be present in the reference phase function $\gamma(x,y)$.

The electrical signal generated by the image dissector is proportional to the light intensity in the x,y plane given by

$$I(x,y) = |U(x,y)|^2 = 2 + 2\cos\{kA(x,y)\sin[\omega t + \theta(x,y)] + \beta(x,y) - \gamma(x,y)\} \qquad \textbf{48}$$

89

Typically, the perturbations of the reflecting surface 6 will have small amplitudes as compared to the wavelength of light, that is, $kA(x,y) << 1$. Thus the intensity is well approximated by

$$I(x,y) \simeq 2 + 2\cos(\beta - \gamma) - 2kA\sin(\omega t + \theta)\sin(\beta - \gamma) \qquad \textbf{49}$$

The reference beam 8 deflected by beam splitter 3 is cyclically temporally offset by a modulator to form a modulated reference beam. In the demonstrated example the modulator consists of piezoelectric crystal stack 11 having retroreflector surface 12 affixed at the end stack. Piezoelectric crystal stack 11 is driven by sawtooth voltage generator 13, having a substantially linear ramp and fast recovery which returns from a maximum to zero every $1/r$ seconds where r is the repetition rate of the sawtooth wave. The amplitude of the sawtooth voltage is adjusted so as to cause the reflected reference light wave to suffer a total phase advance or retardation of n 2π during each sawtooth cycle, where n is an integer. The resulting motion of reflecting surface 12 results in a continuous linear-with-time phase shift of reference beam 8, causing an effective frequency modulation of the reference beam. As a result, the complex interference pattern is superimposed on a carrier and the entire pattern displayed on the face of image dissector 10.

Image dissector 10 scans the complex light pattern and generates an analog electric signal having components corresponding to the desired image information obtained from the sound image and, in addition, other components corresponding to spurious vibrations and undiffracted waves. The important feature of this technique is that the resulting electric signal is amplified and passed through narrow band filter circuit in order to extract the signal components corresponding to the sound image to be visualized. The filtered signal is thereafter demodulated and displayed on an electronically addressable optical display, e.g. cathode ray tube or solid state display.

4.3.4 Electronic sampling

The visualization of sound images via electronical sampling is a relative to mechanical scanning because in both cases sound image formation is based on sampling, only in electronical scanning more than one sampling aperture is used. Two distinct modes of use can be distinguished, but in both cases the electrical properties of the aperture elements are utilized as signal sources for the visualization. The difference is in the mode of interrogation. In the first case the sampling of the sound image formation is achieved by working successively through the complete series of apertures or groups in a 'one element a time' form. In the second, more sophisticated approach the scanning process involves steering of the sound beam issuing from the aperture by appropriately varying the individual time delays introduced between the various apertures.

4.3.4.1 Sampling apertures in two-dimensional array In first approximation this sampling strategy can be regarded as a version of sampling sound images projected on an area detector surface. To place things in proper perspective, however, a few rules of thumb have to be mentioned concerning the design of sound image visualization via this sampling strategy.

First of all, the size of the sound image to be visualized is a critical factor and is generally limited not really by physical but rather by economical constraints. A fair compromise for the lower limit of a sound image diameter seems to be 100λ. Assuming a square array of apertures with an area of a picture element, a pixel, a reasonable upper limit would be a 100 by 100 point array, i.e. 10^4 transducers. Since the angular resolution α for a square transducer is given by

$$\sin \alpha = \lambda/A \qquad\qquad\qquad 50$$

where A is the pixel area, and the half angle field of view β is given by

$$\sin \beta = \lambda/2d \qquad\qquad\qquad 51$$

where d is the spacing between the transducers, the relation between α and β is given by

$$\sin \beta = 50\sin\alpha \qquad\qquad\qquad 52$$

if 100 by 100 array is considered. Since

$$\beta = \sin^{-1}(50\alpha) \qquad\qquad\qquad 53$$

and since

$$\sin \alpha \sim \alpha$$

the number of resolvable picture lines becomes

$$\text{total lines} = 2\beta/\alpha = 2[\sin^{-1}(50\alpha)]/\alpha \qquad\qquad\qquad 54$$

The choice of the sampling element material varies from the sound imaging problems to be solved, but at present the assortment is not too large. Naturally, the first consideration is a piezoelectric material, quartz or ceramics. The usual technique for making such piezoelectric array is to bind a ceramic material to a glossy backing and cut slots between the individual elements[136]. Sometimes the individual elements are assembled separately on the backing[137].

An alternative approach would be to prepare the sampling array in such a way that the individual elements of the array are defined only by deposition of metal films on a ceramic. Standard photolithographic technique could then be used instead of the slotting method. Unfortunately, however, the sound wave entering such an array from the coupling medium, generally water, will be reflected and re-reflected within the ceramic, because of the high reflection coefficient of waves at each surface of the ceramic. These reflections add up and give a strong signal near normal incidence. Information from, for example, specular reflecting points gets lost because at angles other than normal incidence the amplitude falls off radically. Further, there will be very little isolation between neighbouring elements, which is naturally not the case if the backing is slotted.

To eliminate these difficulties Auld et al[138] suggested making a good acoustic match to either one or both the back and front faces of the sampling area. In this case the Q of the system will be small, and an entering plane wave will not be reflected from the matched face. Thus, there is no build up of field of normal incidence, and the angle of acceptance will approach the critical angle for the medium, e.g. for lead metaniobate about 32°. Theoretically this requirement can be fulfilled by using λ/4 matching sections on the front side, and having lead or tungsten loaded epoxy backing. In practice, however, it is difficult to make a matched impedance backing which remains matched over wide acceptance angles. Fig. 35 shows the results obtained by Auld et al at 1.2 MHz using a lead backing and λ/4 SF6 glass and plexiglass layers on the front. The acceptance angle is approximately ±15°. This method, when construction technology is improved, may lead to large piezoelectric transducer arrays with good angular response and good isolation between sampling apertures with no slotting required between them.

Piezoelectric plate as sampling aperture in an array has, however, the inherent

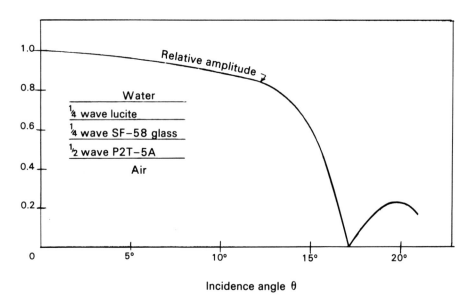

Fig. 35 Acoustic matching after Auld et al[138].

disadvantage that to achieve good threshold sensitivity it has to operate near reso-
nance, which results in small bandwidth. An array consisting of condenser microphones would eliminate this problem, but, the conventional condenser microphone
consisting of an edge supported metal diaphragm is not suited for use at those sound
frequencies which are needed for enjoyable sound images. Moreover, a simple technique for incorporating it in a large array of similar units is not presently known.
However, using electrostatic microphones of the foil electret type[139] we do not
have to face these problems.

The electret microphone as shown in Fig. 36 utilizes a thin polymer foil metallized
on one side and permanently charged on its polymer surface. The charged foil is

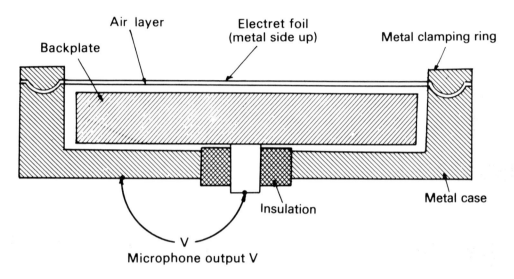

Fig. 36   Basic layout of an electret microphone.

stretched across a metallic backplate with the metal side facing out. Due to microscopic irregularities in their surfaces, the backplate and the polymer diaphragm
physically contact each other at a large number of points, which makes it possible to
operate such a microphone up to frequencies of 1–10 MHz. Thus, as shown by Nigam
et al[140] if a multielement subdivided backplate is used, a large area electret microphone inherently acts as an array of microphone elements. It is termed 'N^2 array'
and can be sampled in parallel or in sequential format. Although the possibility of
the sequential sampling scheme has been proved experimentally, it will be econom-
ically feasible only with further development of the integrated circuit and large scale
integrating technique whereby the switch matrix for a large array can be directly
integrated at the backplate.

Since parallel readout is not frequently needed, the number of subdivisions of the
backplate and also the number of switches can be reduced considerably by subdividing

both the backplate and the metal layer of the foil along narrow strips as shown in Fig. 37[141]. Each foil strip partly overlaps all backplate strips and each overlap

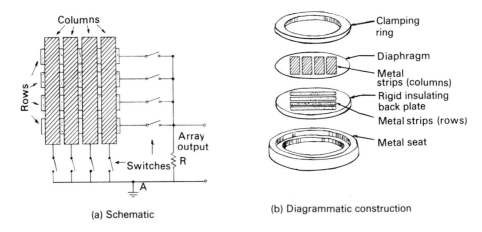

(a) Schematic

(b) Diagrammatic construction

Fig. 37   The basic layout of a 2N electret array.

forms a separate microphone. This means that for an N x N element array both backplate and foil have N subdivisions each and, therefore, it is called by the inventors a '2N array'[142,143].

Both types of array can be optimized for operation in any particular medium over certain frequency ranges. Theoretically threshold intensity at unit signal-to-noise ratio of 10^{-8} W/cm^2 in air, and 2×10^{-11} W/cm^2 in water can be obtained at 1 MHz. The value in water is comparable to the minimum detected intensity by piezoelectric arrays. Parameters which affect the performance of these arrays are variations in sensitivity across the array and crosstalk between the elements of the array.

Reasons for the variation in sensitivity are beside nonuniformities of charge density on the foil, the geometric nonuniformities such as those in air gap and element size, a further nonuniformity in closed-switch resistances.

Crosstalk between elements may arise from finite mechanical transfer impedance of the stored foil, but the electrical coupling between backplate elements and among switches is more important. Experiments have shown that electrical crosstalk for the externally biased 2N design is of the same order as for the N^2 design, whereas the electret biased 2N design shows considerably higher crosstalk.

4.3.4.2 Sampling apertures in one-dimensional array   The only advantage of sampling sound images with two-dimensional aperture arrays is that the image points can be processed in parallel, but using somewhat more sophisticated electronically scanned one-dimensional aperture array the same real-time visualization of the sound image can be obtained. Two distinct modes of solution can be distinguished.

In the first case, the time required to obtain one sound image of B-mode type is

94

determined by the number of elements switched and the pulse recurrence, with a minimum of 30 total visualized sound images required to avoid flicker (Fig. 38).

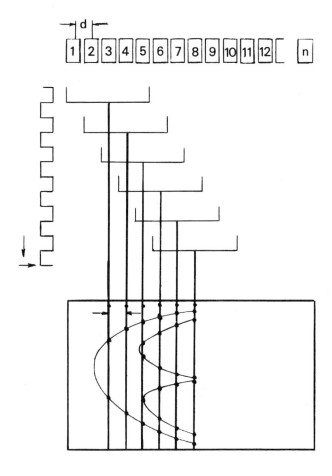

Fig. 38 Method of linear electronic scanning.

Since pulse repetition rate equals the time from ultrasonic pulse generation until the reflected echo vanished, it cannot be made infinitely short. One of the ways to solve this problem is illustrated in Fig. 39. The three elements located at the end of the array are initially used to generate the sound pulses, and when echoes from the farthest point of the target space are sampled, three elements located in the middle of the array are used to send ultrasonic pulses in such a way that crosstalk cannot occur. Then groups of elements separated by a given distance operate reciprocally in sequence, with the change-over frequency set at the minimum necessary value.

95

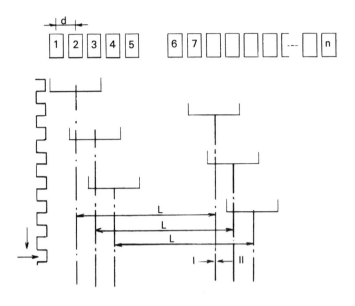

Fig. 39 Pulse frequency shortening method.

In order to solve the problems of reducing the number of sampling apertures to decrease the number of changes over time without increasing the scanning space the so-called interlace method was introduced. As demonstrated in Fig. 40, in this method the number of simultaneously operated sample apertures is reduced, but instead of being constant it is increased in sequence, which causes the center of the ultrasonic beam to be displaced half the distance occupied by the elements[143].

Defining phase as a function of element position to direct the trajectories of sound pulses, i.e. to 'steer the sound beam', and to accomplish thereby a visualization of sound images is well known in radar technique[144] and was adopted to underwater acoustics in the early 1960s[145], and to ultrasonic medical diagnostics at the beginning of 1970s[146, 147].

This technique is sometimes termed a phase array method, but since in ultrasound pulse-echo technique pulses as short as possible are used, it is perhaps more accurate if we speak of time delay rather than of phase differences. The corresponding maxima and minima of the short pulses can be used as reference points to each of which a corresponding wavefront is ordered. If the time delays of the corresponding wavefronts are: id sin θ/c (c is the sound velocity, d is the distance between elements) with respect to the reference element, then the direction of the main beam will be at angle θ with the normal as will be clear from Fig. 41. By changing the main control voltage on which sin θ depends linearly, the direction of the main beam can be

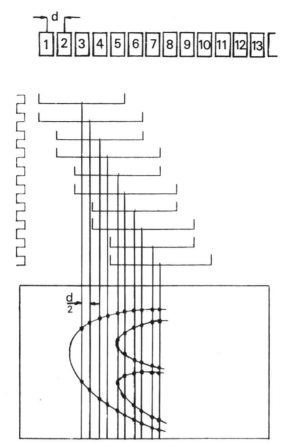

Fig. 40   Interlace scanning method.

varied. If the elements are not in a line but circularly arranged, the sound beam can be rotated around the axis as demonstrated by Bom *et al*[148] who developed an ultrasonic intercardiac scanner on this basis. Fig. 42 shows two consecutive scanning positions on this ultrasound catheter tip in the receptive mode. The directivity pattern is formed using a four-element combination.

While in transmission, i.e. to insonify the space to be visualized, the steering of the beam is relatively easy, in reception, the respective delay line and its change over is quite complicated. Several technical solutions exist and the reader interested in this most promising new method of seeing by sound is referred to an ever growing number of references[149, 150, 151].

97

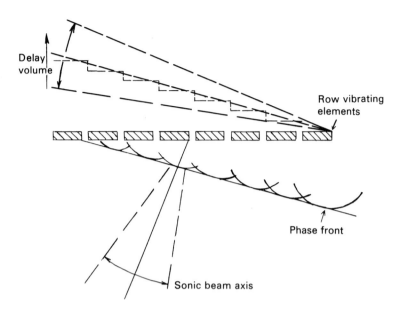

Fig. 41 Deflection by phase control.

4.3.5 Acoustic lens sampling

To use an acoustic lens to form a sound image is in reality a sampling procedure which shows the general relationship between an object space, its spatial spectrum, the imaging aperture and the image[152] (Fig. 43).

The aperture $G(x,y)$ is sampling a portion of the spatial frequency spectrum of the target point distribution $S(x,y)$. If the complex amplitude distribution over the aperture undergoes a Fourier transformation (e.g. using a lens), the result is again the image $O(x',y')$. In other words, the sound image $O(x',y')$ is equal to the convolution of the directivity pattern of the aperture $G(x,y)$ and the distribution of the target points $S(x,y)$.

The position of the aperture and the sound source (target points) determined the spatial frequency components of the scene available at the aperture, i.e. the position and the size of the aperture in the spatial frequency domain determines the position of the spatial spectrum, the target points that can be imaged. Fig. 44 is a plot of the relationship between image and target distance and the focal length of the lens being the aperture. Considering the shaded area into the lower part of Fig. 44, it can be seen that for target distances greater than twice the focal length the image is formed somewhere between one and two times the focal length in the back of the lens.

Another description of forming a sound image with an acoustic lens is to consider

98

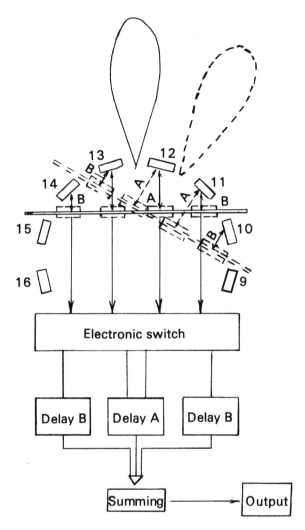

Fig. 42 Two consecutive scanning positions on an ultrasound catheter tip in receptive mode. The directivity pattern is formed by using four-element combination.

that the energy travelling along each raypath through the lens must arrive simultaneously at the corresponding focal point, i.e. each path must have the same transit time between object and focal point. In a practical lens, however, the various raypaths do not converge. Fig. 45 shows raypath diagrams through an acoustic lens for a point on the axis and for one 20° off axis, where aberration becomes noticeable. There are various forms of aberrations, such as spherical aberration, aberration of the field curvature, etc.

The various aberrations can be minimized but never simultaneously eliminated in

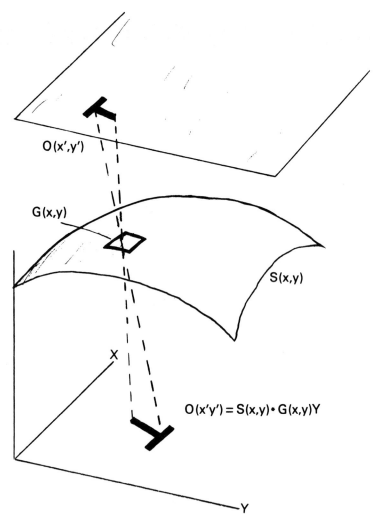

O(x',y')

G(x,y)

S(x,y)

X

$$O(x'y') = S(x,y) \cdot G(x,y)Y$$

Y

Fig. 43   Relation between an object space, its spatial spectrum, the imaging aperture and the image.

practical lens design. Although the first acoustic lenses for sound imaging were already made in the late 1930s, there are still no lens catalogs from which so-called standard designs may be chosen to meet the specific requirements. However, in the past few years interdisciplinary research has been performed in this field, and the reader who is interested to explore sonic imaging design which could approach photographic quality in visual display may find good hints in the references[153–160], while here we are listing only the basic guidelines for the design of imaging acoustic lenses.

100

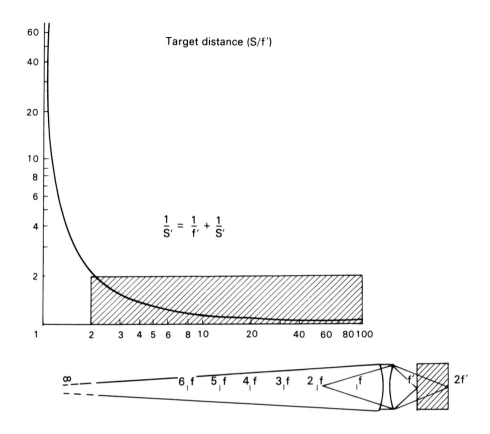

Fig. 44 Relationship between image and target distance.

a) The minimum object distance is an important consideration in choosing the focal length of the lens. Short focal length means higher gain of the lens, and a better resolving power, as shown in Fig. 46, for relative aperture of $f\,1$ and $f\,2$, where D is the lens diameter, and λ the wavelength. On the right hand, as focal length is increased, the off-axis aberrations are reduced, the depth of focus improved.

b) The distance of two target points which must be resolved by the lens, control the design of a practical system too. It would be advantageous if the wavelength could be chosen very short as compared to the distance to be resolved and *extremely* short as compared to lens diameter. However, the frequency must be as low as possible in order to reduce absorption and scattering losses.

c) The focal length of the lens should be independent from temperature over a wide range. Such an athermal doublet lens can be fabricated from a material having low sound velocity and from a material having high sound velocity. However, the focal length of a single refractive index thin solid lens made of materials such as

101

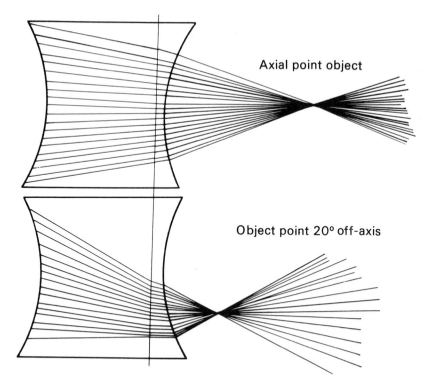

Fig. 45 Raypath diagrams through an acoustic lens for a point on the axis and for one 20° off axis.

polystyrene or synthetic foam is also virtually temperature independent.

d) Due to Rayleigh's criterion a lens cannot concentrate wave energy into an arbitrarily small region. Instead, the energy will be spread by deflection at the focus into a series of concentric rings and this distribution will be further affected by the geometric aberrations. It is quite a good acoustic lens which can concentrate more than 60% of the energy in the central zone.

As shown in Fig. 47, generally the following term is understood by resolution of the lens:

$$\delta = 1.22 \, \lambda/\sin \varphi \qquad\qquad\qquad \textbf{55}$$

where λ is the wavelength in the image space, and φ is the angle subtended by the lens aperture from the focal point.

The crispness of the sound image can be improved by 'stopping down' the lens, since rim rays, contributing most heavily to aberrations, are blocked out, although this increases diffraction. In every case, a compromise has to be found experimentally, taking into account that images formed with middle ultrasonic frequencies will have predominant diffraction characteristics due to the long wavelength. As a conse-

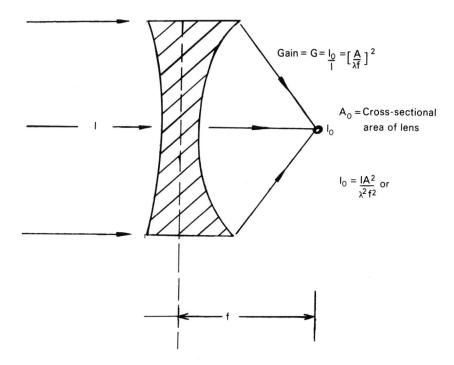

Gain $= G = \frac{I_0}{I} = [\frac{A}{\lambda f}]^2$

$A_0 =$ Cross-sectional area of lens

$I_0 = \frac{IA^2}{\lambda^2 f^2}$ or

Fig. 46 Short focal length means higher gain of the lens and a better resolving power.

quence, scattering points in the image are ringed by holes, the disturbing effect of which can be reduced by properly stopped down acoustic lenses.

e) If large field of view, up to 100°, is needed, single refractive index liquid lenses should be considered, taking, however, into consideration that spherical aberrations limit these lenses to angular resolution not better than 0.5–0.8° in contrast to solid lenses which may have 0.2° with an *f*/number near unity.

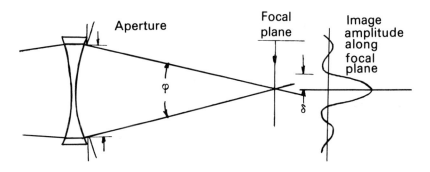

Fig. 47 Resolution geometry of an acoustic lens.

103

f) A really high angular resolution can be achieved by using multielement solid lenses, or the combination of solid and liquid lenses.

4.3.5.1 Electronically focused lenses Sound images, as with optical images, can be formed not only with lenses but also with Fresnel zone plates. The acoustic zone plates were first investigated by Pohlman in the late 1930s[161], and recently by Chao *et al*[162].

The elementary properties of Fresnel zone plates are well described in the literature[163], but are summarized here for convenience.

As shown in Fig. 48, the edge of the nth zone is located at radius r_n, given by

$$r_n = r_1 \sqrt{n} \qquad n = 1, 2, \ldots \qquad\qquad\qquad\textbf{56}$$

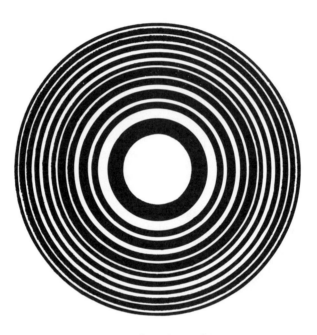

Fig. 49   Fresnel zone plate.

The zone plate may be either positive (center transparent) or negative (center opaque).

If a plane wavefront of wavelength λ is incident on the zone plate, each phase factor will modulate the phase of the wave by an amount proportional to r^2, but leave the amplitude unchanged. The wave which emerges from the zone plate is then a superimposition of the spherical waves, one for each term in the summation, plus the undiffracted plane wave or dc term.

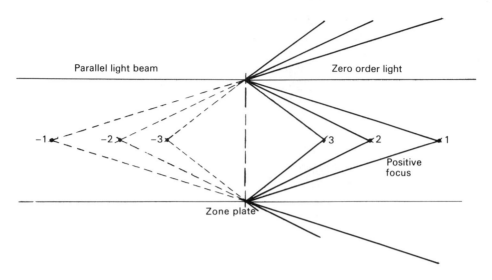

Fig. 49   Fresnel zone plate acting as both positive and negative lens.

As shown in Fig. 49, converging spherical waves will focus after propagating a distance f_p given by

$$f_p = r_1^2/\lambda p \tag{57}$$

and the diverging waves appear to emanate from virtual foci located behind the plate at a distance f_p also given by Equation **57**. Thus the zone plate acts as an infinite series of both positive and negative lenses. The higher orders $|p| \geq 1$ result from the sharp edges of the zones, which produce higher spatial harmonics.

The analogy with a lens is a direct mathematical one. This formal identity between a lens and any one term in the zone plate transparency function permits us to use all of the well known results of the properties of diffraction limited lenses. Thus for a Fresnel zone plate with N zones the diameter of the focal spot on the Rayleigh criterion is given by

$$\delta = \beta\lambda f/D \tag{58}$$

where β is a numerical factor having the value 1.22 for a lens (Equation **55**). For a zone plate, β is generally a function of the numbers of zones N, and depends on whether the zone plate is positive or negative[164]. It approaches 1.22 as N approaches infinity, and is within 10% of this value for N $\geq$ 7.

If the depth of focus is defined as a displacement from the focal plane which gives a 20% reduction in intensity at the center of the focal point

$$\delta_z = \beta\lambda(f/D)^2 \tag{59}$$

where $\beta \approx 2$, but in practice 3 is more appropriate since centre intensity reduction of 50% should be considered.

Sound imaging with Fresnel zone plates can be passive or active. The sound imaging is considered as passive if the zone plate is made of a sequence of alternatively transparent and opaque zones, the radii are being chosen according Equation **59** and the zone plate is used as a normal acoustic lens. In order to realize active acoustic zone plate it is necessary to excite a thin-plate acoustic transducer only in the region corresponding to the transmission zones of the zone plate pattern. This can be accomplished as demonstrated by Farnow and Auld[165] by using photolithographic techniques to evaporate, e.g. a gold 10–zone negative zone plate pattern as an array of electrodes on one face of the ceramic transducer, as illustrated in Fig. 50. The opposite face is a full-face electrode. Thus in applying an electric field to the transducer only the transmission zones are 'active'. This means that the same zone plate can be used for insonifying the target and for imaging it.

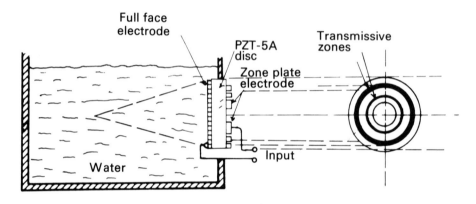

Fig. 50   The acoustic zone plate transducer after Farnow and Aould[175].

A very interesting and most promising feature of the active acoustic Fresnel zone plate imaging is that if the gold zone plate electrode system is replaced with a photoconductive layer sandwiched between the transducer and the optical transparent electrode, it can be switched on with light[166]. By displacing the optical zone plate pattern projected on the photoconductive layer, the acoustic focal region also shifts and this allows one to recognize a scanned zone plate system.

A disadvantage of the acoustic zone plate is that it responds to an incident plane wave. This problem can, however, be overcome if the piezoelectric ceramic is poled according to the alternating of the zones, i.e. a binary phase reversal acoustic zone plate is made[167, 168].

In 4.3.4.2, we have already discussed the feasibility of steering an ultrasound beam produced by a row of piezoelectric transducers. Since the steering of the acoustic beam is a result of an electronic program by which they are excited, it goes without saying that having an appropriate program, i.e. by giving appropriate delays to the signals of the various elements a wavefront equivalent to that of a concave focused transducer can be obtained as demonstrated in Fig. 51.

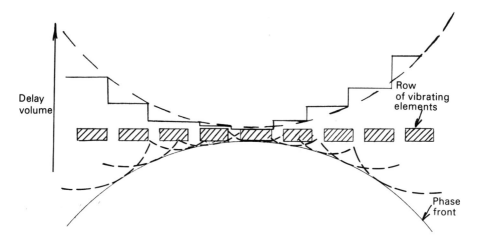

Fig. 51   Concept of electronic focusing.

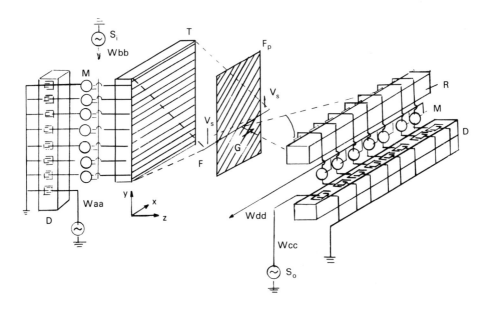

Fig. 52 Schematic pictorial diagram of an electronically scanned and electronically focused acoustic imaging device after Kino *et al*[170].

107

Based on this concept electronically focused sound imaging can be accomplished[169, 170]. The principles of such a system can be well understood from the schematics of Fig. 52. The transmitter consists of an array of piezoelectric transducers which are excited from taps on an acoustic surface delay line, but it is also possible to employ a shift register in a very similar system. The transmitter emits a beam focused and scanned in direction y, acting like a cylindrical lens moving with velocity v_y in direction y producing a beam focused to a line in a plane at distance z from the array.

The receiver is focused on the same plane, but is scanned at velocity v_y in direction y, acting also as a cylindrical lens, focused in direction x and moving with velocity v_x in direction x. The combination of the receiver and transmitter cylindrical lenses focuses on a spot which moves along a line and a plane with velocities v_x, v_y. Since the scan rate in either direction can be arbitrary, the system can operate as a fixed focus lens. Naturally, the system can operate not only in transmission but also in reflection mode, but the electronics needed is somewhat more sophisticated.

References

[44]ARKHANGELSKII, M. E. (1963) Akust. Zh. 9 1–4.
[45]KECK, G. (1959) Acustica 9 79–85.
[46]WEISSLER, A. (1949) Naval Research Lab. Report S3483
[47]GREGUSS, P. (1951) Magy. Kemiai Folyoirat 57 257–263
[48]GREGUSS, P. (1963) International Chem. Eng. 3 280–294
[49]RUST, H. H. (1952) Angew. Chem. 6 308–311
[50]TAMAS, GY. (1958) Magy. Fiz. Folyoirat 6 53–57
[51]BERGER, H., KRASKA, I. R. (1962) J. Acoust. Soc. Am. 34 518–522
[52]OLSON, A. R., GARDEN, N. B. (1932) J. Am. Chem. Soc. 54 791–801
[53]LIU, S. C., WU, H. (1932) J. Am. Chem. Soc. 54 3617–3622
[54]GREGUSS, P. (1956) Szonokemia Mernöktovabbkepzö Intezet, Budapest
[55]HERBIG, J. A. (1967) Encyclopedia of Chemical Technology ed. K. Othmer, Vol. 13 pp. 436–456 John Wiley and Sons, New York
[56]ERNST, P., HOFFMAN, C. (1952) J. Acoust. Soc. Am. 24 207–211
[57]RICHARDS, W. T., LOOMIS, A. L. (1927) J. Am. Chem. Soc. 49 3086–3110
[58]EGERT, J. (1955) Fundamental Mechanism of Photography Sensitivity
[59]GREGUSS, P. (1953) Meres es Automatika 1 34–38
[60]HARVEY, E. N. (1939) J. Am. Chem. Soc. 61 2392–2398
[61]SPENGLER, G. (1953) Nachrichtentechnik 3 399–402

[62]GREGUSS, P. (1978) Colloquium at the Westinghouse Research Laboratory, Pittsburgh
[63]MAILER, H., McMASTER, R. C., GOLIS, M. J., PRIOR, R. B., LIKINS, K. L. Contract No. 0140--690146 ARPA Program Code No. 8D10 (1969)
[64]HARMAN, G. G. (1958) Phys. Rev. 111 27–29
[65]WEISZBURG, J., GREGUSS, P. (1959) Acustica 9 183–186
[66]HAVLICE, J. F. (1969) Electronics Letters 7 477–479
[67]COOK, B. D., WERCHAN, R. (1969) 78th Meeting of the Acoustical Society of America
[68]INTLEKOFER, M. J., AUTH, D. C. (1972) Appl. Phys. 20 151–152
[69]GABOR, D. (1969) Acoustical Holography Vol. 1 ed. A. F. Metherell pp. 267–275 Plenum Press, New York
[70]BÄR, R. (1938) Helv. Phys. Acta 6 570–580
[71]VOSAHLO, H. (1958) Jenaer Jahrbuch 171–178
[72]BABOROVSKY, V. M., MARSH, D. M., SLATER, E. A. (1973) Non-Destructive Testing 6 200–207
[73]BABOROVSKY, V. M., MARSH, D. M. (1971) Ultrasonics 9 86–88
[74]WYATT, R. C. (1971) Central Electricity Generating Board Report No. SSD/SW/M388
[75]WYATT, R. C. (1972) Nondestructive Testing 5 354–358
[76]WYATT, R. C. (1973) Central Electricity Generating Board Report No. SSD/SW/M552
[77]ZOHER, H. (1927) Z. Physik. 28 790–793
[78]MAILER, H., LIKINS, K. L., GOLIS, M. J., McMASTER, R. C. (1970) Contract No.

N00140–70–C–0318 ARPA (1970)
[79]KESSLER, L. W., SAWYER, S. P. (1970) *Appl. Phys. Letters* **17** 440–441
[80]GREGUSS, P. (1973) *Acustica* **29** 53–57
[81]GREGUSS, P. (1973) *Laser and Unconventional Optics Journal* No. 45 1–28
[82]U.S. Pat. 3,831,434 (1972) P. Greguss
[83]GREGUSS, P. (1951) *Magyar Kemiai Folyoirat* **57** 257–263
[84]U.S. Pat. 3,707,323 (1972) L. W. Kessler
[85]POHLMAN, R. (1939) *Z. Phys.* **113** 697–709
[86]VANVALKENBURG, H. E. (1967) 74th Meeting of the Acoustical Society of America
[87]LAFERTY, A. J., STEPHENS, R. W. B. (1972) *J. de Physique Colloque Paul Langevin sur les Ultrasons* **30** 11–12
[88]FINTELMANN, D. Z. (1958) *Univ. Techn. Math. Naturwiss.* **8** 303–310
[89]SETTE, D. (1949) *J. Acoust. Soc. Am.* **21** 375–381
[90]SCHUSTER, K. (1951) *Jenaer Jahrbuch* 217–228
[91]SCHERG, CH. (1975) *Optische Rekonstruktion akustischer Interferenzfelder* MA Thesis, I. Physikalisches Institut der Technischen Hochschule Darmstadt
[92]SPENGLER, G. (1959) *Feingeratetechnik* **8** 406–413
[93]FOX, M. D. (1975) *Appl. Opt.* **14** 1476–1477
[94]BRILLOUIN, L. (1922) *Ann. Phys. (Paris)* **17** 88–92
[95]DEBYE, P., SEARS, W. (1962) *Proc. Nat. Acad. Sci. (USA)* **18** 409–418
[96]KORPEL, A. (1966) *Appl. Phys. Letters* **9** 425–427
[97]KORPEL, A. (1968) *IEEE Trans. Sonics and Ultrasonics* **SU–15** 153–157
[98]KORPEL, A. (1969) *Optical Imaging of Ultrasonic Fields by Acoustic Bragg Diffraction* Drukkerij Bronder-Offset N.V., Rotterdam
[99]SCHLUSSLER, L., WADE, G. (1976) *Appl. Phys. Letters* **28** 695–697
[100]WINTER, D. C. (1973) *Appl. Phys. Letters* **22** 151–152
[101]SOO-CHANG-PEI, WADE, G. (1975) *Int. Opt. Comp. Conf. Digest of Papers* 124–128
[102]GUINOT, J. C., BAERD, B., BILLARD, M., JEAN, R. (1977) 9th International Congress of Acoustics, Madrid
[103]SATO, T., UEDA, M., TADA, H. (1972) *J. Opt. Soc. Am.* **26** 668–671
[104]SATO, T., UEDA, M., (1974) *Ultrasonics* **12** 16–21
[105]MONTGOMERY, W.D., (1969) *Institute for Defense Analysis Research Paper* P-543
[106]HILDEBRAND, B. P., BRENDEN, B. B. (1972) *An Introduction to Acoustical Holography* Plenum Press, New York
[107]MIDDLETON, D. (1960) *An Introduction to Statistical Communications* McGraw-Hill, New York
[108]KORPEL, A., DESMARES, P. (1969) *J. Acoust. Soc. Am.* **45** 881–884
[109]WHITMAN, R. L., KORPEL, A. (1969) *Appl. Opt.* **8** 1567–1576
[110]KORPEL, A., KESSLER, L. W. (1971) *Acoustical Holography* Vol. 3 ed. A. F. Metherell 23–24 Plenum Press, New York
[111]ALES, G. TENNISON, M. A., THOMPSON, R. B., TITTMANN, B. R. *Ultrasonics* **11** 174–177
[112]ADLER, R., KORPEL, A., DESMARES, P. (1968) *IEEE Trans.* **SU–15**
[113]U.S. Pat. 2,164,123 (1939) S. J. Sokolov
[114]Fr. Pat. 3,163,784 (1957) P. Renaut
[115]U.S. Pat. 3,470,868 (1965) W. E. Krause, R. E. Soldner, H. Kresse
[116]PETZOLDT, J., KRAUSE, W., KRESSE, H., SOLDNER, R. (1970) *IEEE Trans. on Biomed. Eng.* 263–265
[117]SOLDNER, R., HAERTEN, R. (1977) *Electromedica* **45** 102–112
[118]GREGUSS, P. (1971) *Proc. 7th International Congress on Acoustics* 473–476 Budapest
[119]McDICKEN, W. N., BRUFF, K., PATON, J. (1974) *Ultrasonics* **12** 269–272
[120]RETTENMAIER, G. (1974) *Ultrasonics in Medicine* ed. M. deVlieger, D. N. White, W. R. McCready, 199–206 Elsevier, Amsterdam
[121]SHAW, A., PATON, J. S., GREGORY, N. L., WHETLAY, B. J. (1976) *Ultrasonics* **14** 35–40
[122]HOUSTON, A. R. et al. (1977) *Brit. Heart J.* **39** 1071–1075
[123]MATZUK, T., SKOLNICK, M. L. January (1978) Private Communication Pittsburgh
[124]BURCKHARDT, C. B., GRANDCHAMP, P. A., HOFFMANN, H. (1975) *IEEE Trans. Sonics and Ultrasonics* **SU–22** 1–11
[125]*Naval Ordonance Laboratory Reports* No. 3967 (1955) and No. 4032 (1956)
[126]JACOBS, J. E. (1963) *IEEE Trans. Ultrasonics Engineering* **UE–10** No. 2, 83
[127]SMYTH, C. N., POYNTON, F. Y., SAYERS, J. F. (1963) *Proc. of the IEE* 110 No. 1
[128]RADIG, G. (1967) *Ultrasonics* **5** 235–238
[129]SCHATZER, E. (1967) *Ultrasonics* **5** 233–234
[130]JACOBS, J. E. (1968) *IEEE TRANS.* **TR-US** 146–152
[131]TURNER, W. R. (1969) *Vitro Sonovision* Vitro Laboratories, Silverspring
[132]DUBOIS, J. (1969) *Ultrasonics* **7** 191–192
[133]MUELLER, R. K., MAROM, E., FRITZLER, D. (1969) *Appl. Opt.* **8** 1537–1538
[134]BROWN, P. H., RANDALL, P. D., SIVYER, R. F., WARDLEY, J. (1975) *A High Resolution Ultrasonic Image Converter*, EMI Ltd. Central Research Laboratory Report
[135]U.S. Pat. 3,716,826 (1973) P. S. Green
[136]LANCEE, C. T., BOM, N., RIJNDORP, H. H.

109

(1975) *Ultrasonics in Medicine* ed. E. Kazner *et al.* pp. 49–53 Excerpta Medica, American Elsevier, Amsterdam, New York

[137]FARAH, H. R., MAROM, E., MUELLER, R. K. (1970) *Acoustical Holography* Vol. 2 ed. A. F. Metherell 173–188 Plenum Press, New York

[138]AULD, B. A., DeSILETS, C., KINO, G. S. (1974) *IEEE Ultrasonic Symposium Proceedings* **74–CHO 996–ISU** 24–27

[139]SESSLER, G. H. (1963) *J. Acoust. Soc. Am.* **35** 1354–1357

[140]NIGAM, A. K., TAYLOR, K. J., SESSLER, G. M. (1972) *Acoustical Holography* Vol. 4 ed. G. Wade 173–194 Plenum Press, New York

[141]NIGAM, A. K. (1974) *J. Acoust. Soc. Am.* **55** 978–985

[142]NIGAM, A. K., SESSLER, G. M. (1972) *Appl. Phys. Letters* **21** 229–231

[143]HIRAFUKU, S., YOSIKAWA, Y. (1977) *J.E.E.* 50–55 May

[144]ALLEN, J. L. (1964) *IEEE Spectrum* **1** 115–119

[145]WELSLY, V. G., DUNN, J. R. (1963) *J. British Institution of Radio Eng.* **26** 205–209

[146]SOMER, J. C. (1968) *Ultrasonics* **6** 153–159

[147]THURSTONE, F. L., VON RAMM, O. T. (1973) *Acoustical Holography* Vol. 5 ed. P. S. Green pp. 249–259 Plenum Press, New York

[148]BOM, N., LANCEE, C. T., vanEGMOND, F. C. (1972) *Ultrasonics* **10** 72–76

[149]PAPPALARDO, M. (1973) *Ultrasonics* **11** 77–82

[150]WEST, R. C., HAROLD, S. O. (1974) *Ultrasonics* **12** 222–226

[151]MILLER, E. B., THURSTONE, F. L. (1977) *J. Acoust. Soc. Am.* **61** 1481–1491

[152]WARNER, H. L. (1972) 84th Meeting of the Acoustical Society of America

[153]TURNER. W. R. (1967) *Vitro Sonovision* Vitro Laboratories

[154]KATZENMEYER, Wm. E. (1970) 79th Meeting of the Acoustical Society of America

[155]FOLDS, D. L. (1971) 82nd Meeting of the Acoustical Society of America

[156]WARNER, H. L. (1972) 84th Meeting of the Acoust. Soc. Am.

[157]U.S. Pat. 3,866,711 (1975) D. L. Folds

[158]U.S. Pat. 3,913,061 (1975) P. S. Green

[159]FOLDS, D. L., HANLIN, J. (1975) *J. Acoust. Soc. Am.* **58** 72–77

[160]SZILARD, J., KIDGER, M. (1976) *Ultrasonics* **14** 268–272

[161]Ned. Pat. OCT ROOI 48400 (1940) R. Pohlman

[162]CHAO, G., AULD, B. A., WINLOW, D. K. *IEEE Ultrasonics Symposium Proceedings* (1972) **72–CHO–708–8SU** 140–141

[163]JENKINS, F. A., WHITE, H. E. (1957) *Fundamentals of Optics* McGraw-Hill, New York

[164]STIGLIANI, D. J., MITTRA, R., SEMONIN, R. G. (1967) *J. Opt. Soc. Am.* **57** 610–614

[165]FARNOW, S. A., AULD, B. A. (1974) *Appl. Phys. Letters* **25** 681–682

[166]ROBERTS, C. G. (1974) Ph. D. Dissertation Stamford University USA

[167]WOOD, R. N. (1957) *Physical Optics* Dover, New York

[168]ALIAS, P., FINK, M., PERRIN, J. (1975) *Ultrasonics in Medicine* ed. E. Kazner et al. pp. 75–80 Excerpta Medica American Elsevier, Amsterdam, New York

[169]U.S. Pat. 3,875,550 (1975) C. F. Quate, J. F. Havlice, G. S. Kino

[170]U.S. Pat. 3,953,825 (1976) G. S. Kino, C. F. Quate, J. F. Havlice

5 DISPLAYS FOR SOUND IMAGES

The objective of this book is to give the reader the ability to specify the parameters affecting the formation of an acoustic image and the ability to estimate the best mode of display for the visualization of the sound image to accomplish a given task.

The earlier chapters gave a background and solid acquaintance with sound image formation methods and the literature documenting them. Terms such as 'optical replica of the sound image' or 'visualized sound image' were used, but in reality only the 'readout' method of a sound image was discussed, even if this readout resulted in an optical image.

We have to distinguish between the visualized and the displayed sound field, since the purpose of a display is not only to convert the sound image into a luminous message to the retina of the eye, but to create a condition where the insonified scene can be discerned with sufficient clarity to serve some intended function and where the observer is highly motivated to do so. The words "sufficient clarity" should be stressed. In one case it may be sufficient merely to detect a blob such as a foreign body, while in another case much higher acuity is needed. For example, it is no use to display a sound image in 6 or 8 gray scales if our *a priori* knowledge is insufficient to interpret it.

A sound image display, similar to electro-optical displays[171], enables

a) *detection* which is set to occur when the observer correctly indicates his decision that the feature of interest exists in the field of view or,

b) *recognition* which is set to occur when the observer correctly indicates to which category the detected feature belongs to or,

c) *area recognition* which is set to occur when the observer correctly indicates where the location of feature in the field of view is or,

d) *discrimination* which is set to occur when the observer is correct in separating a single feature from the flux of recognized targets *or*

e) the combination of these characteristics.

It has to be emphasized, however, that these characteristics are not considered to grade sound image display, only to indicate that the relationship between the physical dimensions of an input and the observer's perception of this input is also a matter of psychophysics, i.e. designing a display intended for 'seeing by sound', the response characteristics of the human visual system should be included in the performance specification. Further information is given by Poole [172] and Luxemberg [173].

Nevertheless, we have to discuss briefly some of these problems since with the spreading of sound imaging techniques, especially in medical diagnostics, there seems to be a dangerous tendency to identify the visualized and displayed sound image with the insonified target based on the *a priori* knowledge of its optical appearance. It is often forgotten that there is a *qualitative* difference between the invisible sound image and its visualized and displayed pattern, while the invisible electromagnetic image differs only *quantitatively* from its visualized and displayed pattern. Therefore all effort to manipulate at the display level in order to obtain a 'visually acceptable' image has to be handled with great care. Differences in the *optical* appearance of a displayed sound image and a displayed invisible elctromagnetic image does not mean that one or the other is 'incorrect', that it does not show 'reality', only that they convey two different messages, which may or may not be related to each other.

Plate 19 and Plate 20 illustrate this point; they are two photographs taken by Szebeni *et al*[174], from livers with suspected malignant involvement. Plate 19 is an

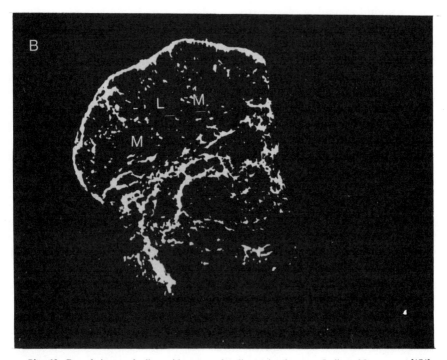

Plate 19   B-mode image of a liver with suspected malignant involvement. L, liver: M, metastase[174].

ultrasonic B-mode image while Plate 20 is a hepatoscintigram of the same liver. The morphological appearance and the spatial appearance (localization) of the lesions are different and they have to be, since they are images of two different information patterns of the same object.

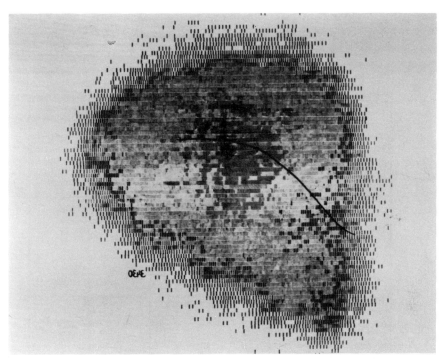

Plate 20 Hepatiscintigram of the liver of Plate 19. Solid line, rib.

The following provides a guide to select a proper display technique for non-sampled and sampled sound images, to weigh the message the displayed pattern can convey, and some of the problem areas of opto- and electro-acoustically sampled sound-imaging display systems in the light of psychophysical cues. Correct results should be attributed to the works of Johnson, LaPorte, Calhoum, Davidson and others[175–177], while any errors belong to the present author.

Detection is the lowest level of display evaluation since it usually means only to spot an object in the insonified space, while the *recognition* of the target means a higher level of display evaluation. It is generally assumed that recognition requires shape information, but in several cases, if other cues are available, it is not needed at all. For example, an A-mode display of an insonified tumor in the eye does not say anything about its shape, nevertheless it can be recognized as a tumor. *Area recognition* and *feature discrimination* are difficult to separate, and it is evident that a higher degree of visual acuity is needed. It is said that an object can be discriminated on the display if its signal-to-noise ratio at the output of the observer's retina after processing and interpretation by the brain is large enough. This means, however, that under the illumination conditions used to view a display, when the eye threshold contrast modulation is 0.04, a target image contrast modulation of 0.04 must be realized at the display for the feature to be detected, regardless of the inherent

113

contrast of the insonified target. This is one of the reasons why some of the area detectors for non-sampled images are not suited for practical application.

This criterion is less severe for displays of insonified objects of low spatial frequencies than for targets of higher spatial frequencies, when the effect of the display

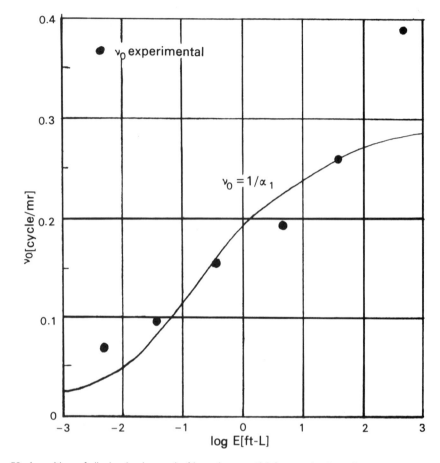

Fig. 53 Logarithm of display luminance in ftL, various spatial frequencies in cycles/mr at minimum noise required input modulation.

noise becomes important. Figure 53 shows the logarithm of display luminance in ftL various spatial frequency in cycles per mr at minimum noise required input modulation.

If we accept Johnson's definition that a spatial frequency of a displayed feature can be specified by numbers of lines in the pattern subtended by the feature minimum dimensions, as illustrated in Figure 54, and for recognition 8 lines with ϵ length to width ratio are needed, we can estimate the feature area/display area ratio, a/A. Let us assume that the feature to be recognized on the display with dimensions XY has

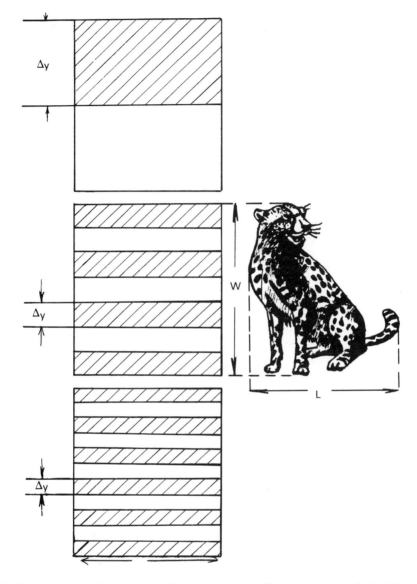

Fig. 54 Resolution required by minimum object dimension to achieve a given level of object discrimination expressed in terms of an equivalent line pattern.

the size of xy. To correlate the detectability of the line pattern with the level of feature discrimination let us divide the minimum dimensions of the feature y by factor k, i.e. $\Delta Y = y/k$. Thus

$$\frac{a}{A} = \frac{x}{X}\frac{y}{Y} = \frac{x\,\Delta y}{k\,\alpha Y^2} = \frac{\epsilon}{k\,\alpha N^2} \qquad\qquad \mathbf{60}$$

where α is a displayed horizontal to vertical feature aspect ratio, and N is the feature's spatial frequency in 'lines per picture height'.

It is apparent from Equation **60** that if a higher level of feature discrimination is desired, higher resolution is needed. Most displays for sound image visualization are at present television-like screens, the line structure of which, although visible, is far better than the Nyquist criterion required for sound image formation. To effect integration by the eye into a structureless field an increased viewing distance is required. However, an image containing a visible line structure may appear to be sharper. Nevertheless, the observer prefers twice the distance for a display with a typical variable raster to that required for an apparently rasterless image[178]. Although more details become visible when the line structure is removed, e.g. by defocusing the CRT, the effect of noise will also be much more visible.

5.1 Sound images visualized in A-mode

The two-dimensional pattern of the screen of an A-mode display does not represent a visualized sound image in the sense of an optical image, since it displays only a single spatial dimension, that of sound propagation direction, which is regarded in general as range and shows up as horizontal or x direction, i.e. as azimuth on the screen. It can also be interpreted as time coordinate, or reciprocal time, i.e. frequency. Direction y, representing in normal optical images the altitude, is displaying in the A-mode form amplitude information, and therefore, our eye-brain combination does not interpret this pattern as an image; we see namely an image only in the spatial domain and do not see a spatial frequency picture. In other words, if the sound field of the insonified object is displayed in such a visualized form, it is difficult, if not impossible, for us to tell something about the dimensions, shape or spatial orientation of the insonified object—the main objective of an 'image'—despite the fact that this information *is* displayed.

The situation is somewhat more favorable if not monochromatic, i.e. single frequency insonification is used but more or less 'white' insonification, i.e. multifrequency wavefront is generated. Krautkrämer[179], was the first who suggested this idea, and Gericke[180] demonstrated qualitatively its validity, while Adler and Whaley[181] developed this idea to determine the shape of arbitrarily oriented two-dimensional objects (e.g. flaws) from A-mode display.

Let us assume that the multifrequency wavefront illuminates under an angle of θ a target having a diameter d and being at a distance D from the transducer. The incident plane wave breaks up into three parts when reflected: a specular reflection part, which, above a small angle, typically 2–4°, misses the face of the transducer, and two sets of wavelets originating from the edges of the target. These wavelets are coherent and interfere at the face of the transducer. Because these wavelets contain a broad range of frequencies, the condition of constructive interference will always be satisfied by some frequencies, as shown in Figure 55. It can be shown that the size

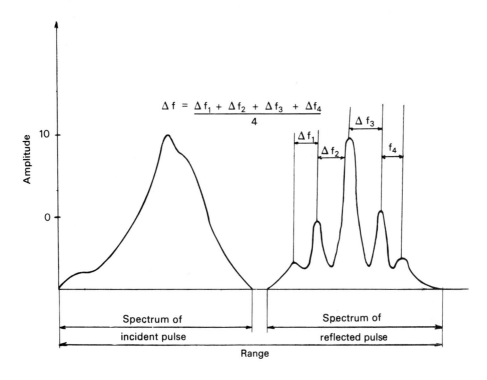

$$\Delta f = \frac{\Delta f_1 + \Delta f_2 + \Delta f_3 + \Delta f_4}{4}$$

Fig. 55 Reflected frequency spectrum showing typical constructive interference.

of the target can be derived, can be 'seen', from the average frequency interval between points of maxima in the spectrum. This is according to Equation **61**

$$d = \frac{c}{(2\sin\theta + a/D)\Delta f} \qquad\qquad \textbf{61}$$

where c is the sound velocity of the medium in which the insonification takes place.

5.1.1 Scanned deflection modulated display

Even when multifrequency insonification is used, a single A-mode display allows only a rather restricted 'image information' from the target, and the angle of insonification has to be known also. More shape identification, i.e. image-like display can be obtained if, not a single, but several A-mode sound images are simultaneously displayed on a screen. This idea was introduced by Baum[182]. According to this suggestion the insonification is accomplished stepwise prescribed by a predetermined program, and the trace line of the CRT is also moved in the same manner as shown

117

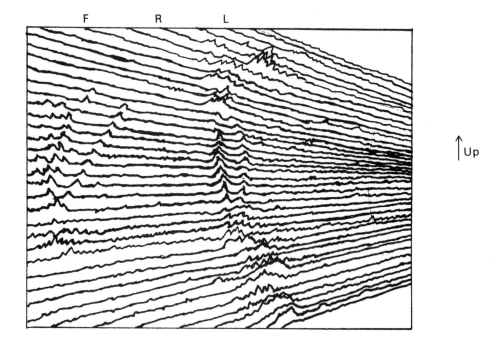

F   R   L

↑Up

Fig. 56   A scanned deflection modulated sound image.

in Figure 56. The result is a pattern providing a real image impression since it contains information on two spatial dimensions beside amplitude information. The unique feature of this display called scanned deflection modulated display is that the A-mode images of adjacent scan lines may be compared, so that alternations in pattern are easily perceived which makes the sound image display form rather attractive, especially if the sudden deflections, marking acoustic boundaries, are further enhanced by increasing the intensity of the deflected electron beam.

5.2   Sound images visualized in B-mode

As long as the pattern visualized in A-mode on the screen is a result of a non-sampled sound image, the pattern visualized by the technique described in the previous section is a result of sound image formed by line or sector scanning program, i.e. it is sampled with the same technique as a B-mode image, only its 'resolution' is inferior. The reason for this is that the 'luminosity' of the insonified target point is represented as a luminous deflection rather than a luminous point, the intensity of which is a function of the sound intensity (amplitude square) of the sound wave reflected from the target point as shown in Figure 57.

118

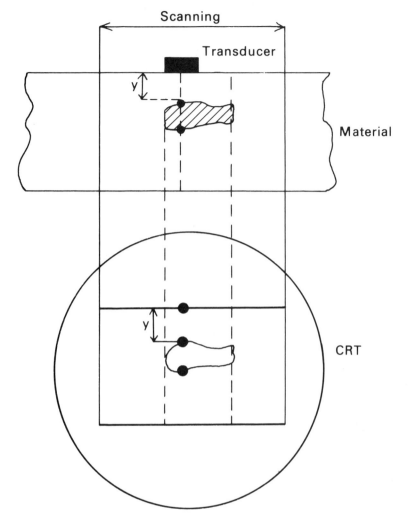

Fig. 57 A scanned intensity modulated sound image.

   In displaying a sound image in the form of a luminous image, the conversion process is continuous in one and discrete in the other spatial coordinate as well as in time. During the formation of a B-mode sound image the transducer yields a set of continuous electrical intensity fluctuation $I_x(t)$ corresponding to a sound intensity fluctuation $I_s(x)$, and this is transformed to a luminance function $L(x)$ along the lines. Luminance function $L(y)$ is, however, the result of the discrete amplitude sample $I_y(t)$ taken at intervals of Δy determined by the raster-line spacing. The displayed B-mode image will show continuity only if the spread S_y of display aperture can fill the

119

entire interline space of the raster. The aperture spread $S(x)$ is generally much smaller, making the resolution in the visualized image anisotropic.

Another possibility to achieve continuity in y is to select a raster line density which provides a large overlap of sampling aperture in successive line scans, but this may be rather doubtful since it may generate aliasing effects, already mentioned in 4.3. In a well-designed display for the visualization of B-mode images these aliasings can, however, be eliminated, for instance, if the first zero of the light intensity from the light spot (see Figure 57) falls on the center of the adjacent light spot. If the screen of a CRT is used as a display, this can be achieved by defocusing the spot of the electron beam so that it covers half of the adjacent scan line.

5.3 Sound images visualized in C-mode

Sound images visualized in C-mode differ from transmission images so that they represent a single plane cross-section of the insonified image space and not cross-sections *compressed* in one plane. In this regard they look similar to the B-mode images. Nevertheless, their information content is somewhat different: namely they represent a cross-section perpendicular to the information carrying wavefront and not the plane parallel to the direction of wave propagation as in the case of B-mode images. C-mode images are obtained by using time-gating and are in general the result of two-dimensional sampling. They can be displayed analogous to B-mode images on the screen of a CRT. However, with the introduction of mosaic of light-emitting diodes, discretely addressable liquid crystal[183], plasma[184], and electro-luminescent[185] mosaic panels, new design possibilities arise.

The signals from each sampling aperture, transducer, after suitable amplification can drive a discrete light source having a spread function $S(x,y)$ at the display. The coordinates of these light sources correspond to the coordinates of the detectors. However, it is appropriate to point out that due to the passive matrix addressing techniques, up till recently such difficulties and limitations arose and research in this direction has been discouraged. However, with the realization of the actives matrices, which perform all the necessary switching, signal distribution and control functions electronically on the displaying screen itself, a new display technique for C-mode sound images could be introduced. It is the author's firm belief that based on the thin film transistor electroluminescent solid state (TFT-EL) display technique developed by Brody *et al*[186], a device could be developed which allows the display of real sound images without the use of acoustic lenses. Plate 21 shows such a TFT-EL solid state display, used as an alpha numeric/graphic display, but there is no reason for why it could not be used as a sound image display. The number of addressable elements is 17 097, it has a resolution of 70 lines/inch, which is more than needed in the frequency range used in nondestructive testing and medical diagnostics. Further, it has 15 levels so that gray scale representation of the sound image would also be possible.

Plate 21   TFT-EL solid state display developed by Brody *et al*[186].

In general, mosaic displays have a rectangular sampling lattice. However, according to 4.3, with a spacing of ½ B in directions *x* and *y* there is a spatial frequency limit B. Nevertheless, the number of detectable line pairs in direction 45° is by a factor of $\sqrt{2}$ greater, i.e. the resolution of such a display is worse in the azimuth and altitude than in direction 45° to the scan direction.

5.4   Sound images visualized in 3-D mode

Visualized sound images represent in general a plane including the line of sight of sound. However, when a space is insonified, the sound waves are scattered basically according to the same geometric principles as light waves are when illuminating the target. As a result, we are faced in both cases with the old frustration that nature is always three-dimensional, but we can record the scattered information from it only in a two-dimensional format, usually losing depth information carried by the phase of the wavefront. The usefulness of the recorded scattered wavefront would be greatly enhanced if it could be reproduced in such a way that we would have not only the feeling of space—as in stereoscopy—but could see spatial features of the scene and could move to look around and behind it, drawing profit of parallax. (Parallax provides the relative motion between near and far features as the viewing eye location is changed, whereas stereoscopy, depth impression, is the result of small differences between the images on the right and left retina.)

Based on the assumption that the perception of three-dimensionality is a result of information on the relative position of two-dimensional layers, or contours of layers[187], attempts have been made to use glass slides of appropriate thickness with B-mode images printed on them to build up the three-dimensional sound space[188]. This method, however, not only suffers from difficulties such as viewing through the edges of the glass and the opacity of information-bearing area, but it does not solve the problem of recording and reconstructing the scattered wavefront from a two-dimensional format either.

The problem is more complicated than one would have thought. Although a monochromatic beam carries n degrees of freedom, the information on it is not as one may think a vector in n dimensions, but a *tensor* with n^2 dimensions. Fortunately, if the beam is coherent, and in acoustic imaging this can be assumed, only $2n$ data are required, since in this case well-defined amplitude and phase are related at every point. Simultaneous recording and/or reconstruction of the information carried by the amplitude and phase of the wavefront can, however, be achieved by the two-step imaging process discovered by Gabor[189], known as holography.

The exact meaning of holography is, however, the subject of discussion. In the original concept of holography, information recording and reconstruction are both wave-optical in nature. The interference pattern between object beam and reference beam is recorded while reconstruction is achieved by diffraction of a reconstruction beam through the recorded pattern, called hologram.

According to the definition of 'generalized holography' which includes what is universally called 'holography' as one subcase and many other additional methods, it is an image encoding and decoding process where encoding involves recording the amplitude and direction of the wave scattered by the object, while decoding involves the *direct* and simultaneous utilization of the amplitude and direction information the recording contains. As a consequence, if the scattering target was three-dimensional, the reconstruction will demonstrate three-dimensional properties, e.g. parallax.

In the definition of generalized holography there is, however, no restriction on the nature of the scattered wavefront. Holograms, therefore, can be recorded not only with electromagnetic but also with mechanical waves or even formed via computer.

Since complete information on the theory and on different techniques of holography are readily available in textbooks[190–195], we shall deal with them so far as it is necessary to judge the value of generalized holography in 3–D.

122

5.4.1 Acoustic holography

A hologram by definition is no more than an interference pattern originating, for instance, from the interaction of the wavefront scattered from the object, $F(x,y)$, and a coherent reference background exp(-kr), and it is obvious that such a pattern can be produced during insonification. If the optical replica of this acoustic interference pattern, i.e. the transparency of it, has an amplitude transmission proportional to the intensity of the two interfering acoustic waves, then, if illuminated with coherent light, one obtains in the plane behind an optical amplitude and phase distribution $H(x,y)$ which contains, beside some other additional terms, a term equal to the phase and amplitude distribution of the object wave $F(x,y)$:

$$H(x,y) = (1 + |F|^2)\exp[i(k_1x+k_2y)] + F(x,y)+F^*(x,y)\exp[i2(k_1x+k_2y)] \qquad 62$$

In the senses of Huygens' principle an optical wave develops behind the transparency, and due to the second term on the right-hand side of Equation 62, a sound image can be visualized in space.

The equation determining the position of the visualized sound image is

$$\frac{1}{\lambda_L D_2} \pm \frac{1}{\lambda_s m^2 D_1} \pm \frac{1}{\lambda_s m^2 R_1} + \frac{1}{\lambda_L R_2} = 0 \qquad 63$$

where λ_s and λ_L are acoustic and laser beam wavelengths, D_1 and D_2 are object and image distances, R_1 and R_2 are the radii of curvature of the reference and reconstruction beams, and m is the size reduction factor introduced in the recording of the hologram.

The reason for introducing a size reduction factor is that due to the difference in wavelength between the insonifying object wave λ_s and the reconstructing light wave λ_L the reconstructed wavefront appears to come from an object field with unchanged lateral (x,y) dimensions, but with depth distortion proportional to λ_s/λ_L. A completely realistic 3–D reconstruction can, however, be obtained only if the lateral and longitudinal magnifications are equal. The lateral magnification can be expressed as

$$M_{lat} = (1/m)\ (\lambda_L/\lambda_s)\ (D_2/D_1) \qquad 64$$

and the longitudinal magnification as

$$M_{long} = (\lambda_s/\lambda_L)M_{lat}^2 \qquad 65$$

A three-dimensional sound image can, therefore, be visualized without distortion only if the lateral magnification equals λ_L/λ_s. In the most common domain of sound

123

imaging, the ratio λ_L/λ_S has a magnitude of about 10^{-3}, thus the undistorted image will be very small and optical magnification has to be applied to obtain useful images, thereby regenerating depth distortion.

Since recorded acoustic holograms have to be scaled down by the ratio of the wavelengths, the visualized sound image does not look three-dimensional because in the majority of cases demagnified acoustic holograms do not present the viewer with an aperture (window) large enough to allow useful parallax. For real three-dimensional impression, a 'window' of at least 10 cm is needed, which means that if the acoustic hologram could be taken at a wavelength of $\lambda_S = 10^{-1}$ cm, and green light of wavelength $\lambda_L = 0.5 \times 10^{-4}$ cm is used for reconstruction, we have to start with an acoustic hologram recording of 200 m (!) which is obviously unrealistic.

Several attempts have been made to minimize this problem. Winter[196] suggested that it would be possible to use demagnified ultrasonic holograms with apertures smaller than the normal interocular separation of about 6 cm by varying the effective eye separation by a set of mirrors, so that the observer's angle of convergence will still be within the range of comfort. Further, by placing a pair of prisms between the acoustic hologram and the observer, the size ratio of space image could favorably be altered because the observer's angle of convergence is correctively changed. It can, however, be shown that the angle necessary to compensate for the first order hologram distortion will vary with object space distance so that with this method only in a pre-selected and small space volume can the disturbance be reduced, and can never be eliminated completely, and so it is doubtful whether this idea really has practical value.

The three-dimensional information is, however, there, even in the more realistic hologram dimensions, but it is available in only one plane at a time, i.e. one depth plane can be brought to focus on a screen or vidicon surface, after the other. This type of imaging is called 'tomographic imaging' and is equivalent to C-mode imaging obtained by time-gating and is also related to B-mode imaging.

The question whether holographic technique or B-mode or C-mode imaging is the answer to the 'seeing by sound' problem depends upon the task to be solved. In this chapter we wish to deal briefly with some of those factors which have to be taken into account for choosing correctly.

Two factors which significantly affect the quality of the reconstructed image are the size and spatial frequency response of the recording medium. The size of the hologram recording surface determines the image resolution and the depth of the field, whereas its spatial frequency response determines the angular field of view and the image intensity response. This means, however, that the method and technique by which the acoustic hologram pattern is converted into the demagnified optical counterpart can be of fundamental importance.

Regarding the size of the recording medium which determines the number of fringes in the Fresnel zone pattern, thus influencing the quality of the visually reconstructed image, methods described in 4.1 and 4.2 should first be considered. As a matter of fact, the first ultrasonic holograms have been recorded on sonosensitized plates, some of which had an area of 100 cm^2 or more. Plate 22 demonstrates the visualized reconstructed sound image of a Guerin type tumor implanted under the skin of a rat[197]. The hologram was recorded in 1965 on a sonosensitized plate. Plate 23 also shows a visualized reconstructed sound image of a tumor in a skin flap, but the acoustic hologram was recorded by Weiss and Holyoke[198] with liquid surface levitation method. Comparing the two visualized reconstructions no significant difference in the visual quality of the two images can be seen. This seems to contradict the statement that the size of the area of the recording medium influences the quality of the visualized reconstructed sound image, since the recording surface of the liquid surface levitation method is significantly smaller than that of the sono-

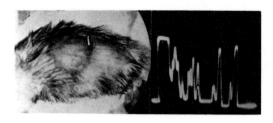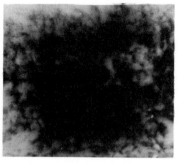

Plate 22 Optical image of a Guerin-type tumor implanted under the skin of a rat, its A-mode sound image, and the image reconstructed from its ultrasonic hologram recorded on sonosensitized plate.

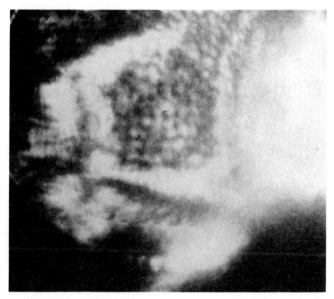

Plate 23 Sound image of a tumor in a skin flap reconstructed from an ultrasonic hologram recorded with liquid surface levitation method[198].

125

sensitive plate. If we consider, however, that the threshold in contrast between the pixels on the recording surface is better if the sound pattern (acoustic hologram) is formed on a liquid surface rather than on a sonosensitized plate, this controversey becomes more understandable.

Unfortunately, the situation is somewhat similar with other non-sampling area detectors too, which can be made in large size, as for instance in the case of AOCC displays. This is all the more regrettable, since they work in real time, and so research to improve their threshold contrast would perhaps be rewarding.

Concerning the importance of the size of the hologram recording medium a Sokolov-type ultrasonic vidicon seems to be a good candidate, since a recording area with a diameter of 100 λ can be taken into account. However, the area of the piezoelectric plate is to be regarded as a closely packed array of receivers with apertures equal to approximately 1.5 times the plate thickness[199], and since the wavelength in the plate is four times larger than in the surrounding water, the angular field of view is restricted to less than 18°. If, however, the piezoelectric plate is replaced by an electret array of the N^2 design, as suggested by Nigam and French[200], and shown in Fig. 58, it appears that angular apertures up to 60° are feasible without appreciable loss in sensitivity. Since, in contrast to the Sokolov-type ultrasonic vidicons with piezoelectric face plate, the thickness of the back of the electret face plate is essen-

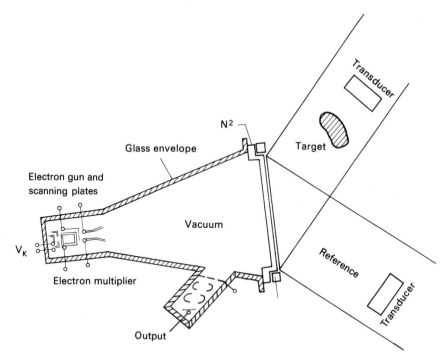

Fig. 58 Schematic cross-sectional view of an electret Sokolov camera (ESC).

126

tially separated from the parameters that determine the sensitivity of the electret elements, physical apertures larger than 100λ can be designed without influencing the uniformity in sensitivity across the face plate.

Looking at the question of spatial frequency, sonosensitized and sonosensitive plates have a better performance than Sokolov-type ultrasonic vidicons, and are equal to liquid surface and solid surface deformation methods, but are somewhat inferior to piezoelectric or similar scanning techniques.

Considering liquid surface deformation methods (4.2.5) which are at present perhaps the most frequently used and advertised techniques for recording acoustic, more precisely, ultrasonic holograms[201–205] we have to keep in mind in the interest of good results that there are three different liquid surface deformations:

a) the deformation occurring at the frequency of the acoustic wave,

b) the deformation due to radiation pressure against gravity, which produces a vertical displacement of the *total* ripple pattern having a natural frequency of about 25 Hz,

c) the deformation which produces the hologram is also due to the radiation pressure, but acts against the surface tension. It has a natural frequency which is proportional to the 3-halves power of the spatial frequency of the ripple, which is generally in the order of 2500 Hz.

The best hologram recording result is achieved when insonification with about $10^{-3} - 3 \times 10^{-4}$ sec ratio is used. Therefore, to visualize the reconstructed acoustic hologram, light flashes in the order of 5×10^{-6} seconds are needed.

Liquid surface deformation techniques to record acoustic holograms are, however, suitable only for the lower ultrasonic frequency range, while solid surface methods can handle frequencies of 100 MHz or even greater, but require a far more sensitive detection and reconstruction technique (4.3.1). So, the method developed by Erikson *et al* [206], uses an optical interferometric technique to record the ultrasonic hologram formed on the mirror surface of a film at motion picture frames, by modulating an *optical* reference beam at the set acoustic frequency. Naturally, reconstruction of these holograms can be performed only after developing the film, which may be regarded as a drawback, but this is compensated by the fact that the viewer may focus any time in any depth within the object under examination. The real problem with this method is that it has a rather high acoustic power requirement which restricts it applications to 1 MHz or so.

The acoustic holographic methods discussed so far record non-sampled acoustic hologram patterns, although in some cases visual reconstruction is the result of a scanning technique. An acoustic hologram pattern can, however, be recorded also by one of the methods by which sampled sound images are formed, (4.3).

There is a great flexibility in the manner in which scanned acoustic holograms can be recorded. One of the simplest methods is when the sampling aperture, a point-like receiver, scans mechanically over the region of object and reference beam intersection, the signal of which modulates a point source in synchrony with the scanning motion of the aperture. The position-intensity history of the lamp, which is in fact the visualized acoustic hologram, is then recorded by a camera. An advantage of this method is that no effective ultrasound reference beam is needed to record the holograms, it being enough to record the scattered ultrasonic field of the insonified object and to simulate the reference wavefront by adding, or multiplying, the received electric signal with a reference electric signal. Multiplication is preferred since it produces the desired signal free of the extra unwanted components that result from the additional process.

Scanned holographic systems are normally operated in reflection mode and use range-gating to eliminate the effect of unwanted reflections. A resolution improvement by a factor of two can be achieved for a given scan aperture size by incorporating the insonifying transducer and the receiving aperture on a time share basis in a single unit. The inherent drawback of this type of acoustic hologram recording is its time-consuming nature, which also prohibits the holographic recording of insonified objects in motion, thus preventing, for instance, some types of medical applications.

A sampling process always introduces some unavoidable data degradation. In the case of mechanical scanning it is simply the actual acoustic excitation that is smeared out over the sampling aperture. In the case of a one- or two-dimensional sampling array, however, the situation becomes more complicated, although the recording time is reduced considerably, nearly to real time. System complexity increases sharply with the number of elements and it is inversely proportional to the square of the sampled interval D. The highest spatial frequency needed to record the acoustic hologram specifies the sampling distance according to

$$D < \lambda S/\sin\Omega$$

wherein $\sin \Omega$ is the field of view of the system. If this inequality is not satisfied, aliasing occurs, that is, ghost images appear in the reconstruction which cannot uniquely be coordinated with a position in the object space.

Although they are quite economical and simple, mechanical scannings of the arrays suffer from the disadvantage of being slow, while electronically interrogated arrays are quite costly at present. The suggestion of Berbekar[207], to substitute mechanical or electrical scanning by sweeping the frequency of the insonifying sound beam may save a number of receivers and also allows practical real-time operation.

This concept is based on the premise that a hologram or the Fourier spectrum of an object contains information on unimportant details[208], and so in most cases of acoustic holography it is enough to pick up only a predetermined part of the hologram, and the number of receivers or processing time can be decreased.

A detector moving along the diffraction pattern with a constant velocity v will experience an intensity variation of

$$I(t) = \frac{k}{\lambda} \left[\frac{\sin(\frac{2\pi d}{2\lambda_0} vt)}{\frac{2\pi d}{Z\lambda_0} vt} \right]^2 \qquad 66$$

where Z is the distance from the slit having a width d, and k is a constant depending on the amplitude of the insonification and the geometry of the arrangement[209].

However, if the detector is fixed and the frequency is varied with the sweeping rate

$$m = \Delta N / \Delta T \qquad\qquad 67$$

where N is the frequency of the wave, the detected intensity as a function of time will be

$$I = \frac{k}{c} \, m \, t \left[\frac{\sin \left(\frac{2\pi d}{cZ} x_0 \, mt \right)}{\frac{2\pi d}{cZ} x_0 \, mt} \right]^2 \qquad\qquad 68$$

In both cases, the sweep of the detected signals will be similar if there is a correlation between the velocity of the detector moving with constant frequency and the speed of frequency sweeps according to

$$v = \frac{m \, x_0}{N_0} \qquad\qquad 69$$

The only difference is that the height of the maximum increases with frequency due to the mt factor before the brackets in Equation 68.

Although by increasing frequency, the distance between the minima in the diffraction pattern decreases, the intensity of the fringes is constant. Therefore it should be compensated for by decreasing the intensity of the insonifying beam in a suitable way. Fortunately, however, this is not even necessary in most cases since there is a 'natural' compensation for it: for most of the materials sound wave attenuation increases with frequency approximately quadratically.

It can be shown that if frequency scanning begins at $t = 0$, and the detector is located at x_0, y_0, due to the compression of the diffraction pattern, the velocity of the virtual detector motion will be

$$v_x = \frac{x_0}{N_0} \cdot \frac{dN}{dt}; \quad v_y = \frac{y_0}{N_0} \cdot \frac{dN}{dt} \qquad\qquad 70$$

and after some manipulation obtain

$$f_x = f_{x_0} + \frac{v_x t N_0}{cZ}; \quad f_y = f_{y_0} + \frac{v_y t N_0}{cZ} \qquad\qquad 71$$

where

$$f_{x_0} = \frac{N_0 x_0}{cZ} \quad \text{and} \quad f_{y_0} = \frac{N_0 y_0}{cZ}$$

are the initial spatial frequencies.

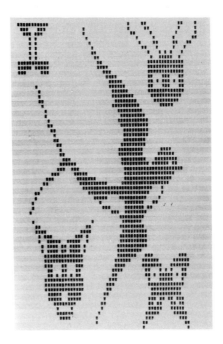

Plate 24 Target for computer simulation[209].

According to Equation **71**, the detector seems to be moving on radii determined by the intersection of the optical axis and the image plane, and by the loci of the detectors. This means that fixed detectors placed on the same radius detect a series of spatial frequencies. The length of the scanned portion depends on the rate of the final and initial frequencies. A great advantage of this method is not only that no moving parts are needed but also that the visualization of the insonified target can be performed with a rather simple computer, for instance, an EC 1010 mini-computer. The validity of Berbekar's idea was tested by computer simulation and gave quite good results. Plate 24 shows the insonified target while Plate 25 is a reconstruction when 64 detectors have been considered.

Since ultrasonic sweep generators and wideband transducers are commercially available, seeing by sound using frequency-swept insonification may have practical value in the future[210, 211).

5.4.1.1 Speckle problem Disregarding the problem issuing from wavelength discrepancy acoustical holography as a counterpart of optical holography has to face another problem, which is essentially not the result of the holographic method itself, but rather the difficult-to-avoid consequence of the fact that insonification with ultrasound means in general a highly coherent 'illumination' of the target. In other words, the amplitude of the scattered coherent wavelets sent into the picture from

130

Plate 25 Computer reconstruction of the target of Plate 24 when 64 detectors have been considered[209].

the volume outside the focus region are summed *vectorially* and then squared. It is not the intensitites that are summed, as if incoherent insonification were to be used, and results in a uniform 'gray' background. This phenomenon called speckle noise

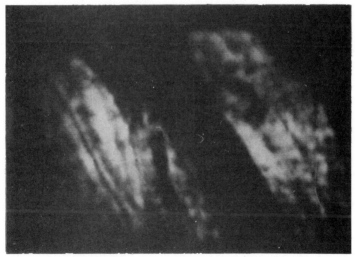

Plate 26 Acoustic image of upper arm showing vascular structures. Liquid surface detector. Frequency scan between 2 and 3 MHz including different 50 frequencies.

131

results in intensity fluctuation which can be as high as 1:10 000, and is responsible for the reduction of resolving power by at least one or two orders of magnitude.

One way to get rid of the disturbing speckles is to average them out by recording several holograms from the same target, insonifying them with reference wavefronts from different angles or with different wavelengths, i.e. with different frequencies, and print them on top of each other. Up until recently, putting this idea into practice was hampered by technical difficulties. Langlois of Holosonics, Inc.[212], succeeded by using a digital storage device, a PEP Scan Converter to store and integrate 50 images, reconstructed from a sequence of acoustic holograms recorded at 50 different frequencies evenly spaced between 2 and 3 MHz. Plate 26 shows such an acoustic image of the upper arm demonstrating vascular structures. The image is free of speckle effects and shows rather well-defined edge contours. It was recorded with surface levitation method using coupling bags. The arm was not immersed in water and so only the region under investigation is imaged, and the edges of the arm are not seen. The image quality is somewhat reduced if real-time visualization is needed, since to get a really flickerfree image 30 frames per second are necessary, so by a framing rate of 100 pictures per second only 5 frequencies ranged from 2 to 3 MHz can be used.

According to Gabor[213], the speckle effect can be avoided if disturbance from layers outside of interest is excluded. This can be accomplished by gating and substituting an acoustic pulse of a duration less than a microsecond for a sinusoidal wavetrain. This pulse when scattered by an object could be displayed on a spherical, thin, metalized, elastic membrane by reflection of high stroboscopic light, preferably a laser. Since the more or less scattered spherical wavelet impinging on the spherical membrane produces a rapidly spreading ring-shaped fine bulge, it will appear in the stroboscopic illumination like a system of Fresnel zones—exactly like the hologram of a point object. Since, for example, the cosine of $15°$ is only by $1/16$ less than unity, by illuminating only during the time in which the fringes spread from zero to $15°$, only $1/32$ of the depth can contribute to the diffraction figure, when operating in an echo-mode. Further, the spread of $15°$ means a f:2, giving a diffraction limit of 2.4 wavelength which is rather good for most nondestructive testing and diagnostic purposes.

This pattern, which is in principle a series of scattered B scans recorded in one shot and can be regarded as a 'faked' hologram, when photographed through a schlieren stop which cuts out the light reflected by the undistorted membrane yields a transparency which, after being processed, can be reconstructed as a hologram with monochromatic light. The idea is simple, but it is not easy to keep a thin membrane in a regular shape so that its zero order reflection can really be cut-out by a small fixed schlieren stop. If a light hologram is recorded—preferably a fraction of a second before insonification—on a photochromic material, using the laser beam reflected from the membrane and a reference beam which has the shape if it would have been reflected from an 'ideal' membrane, then, it can be used as a correcting hologram. Namely, when it is illuminated by the distorted wavefront reflected from the membrane on which the sound field is impinging, it will produce the perfect focus which can be cut-out by a very small fixed schlieren stop.

5.4.2 Holographic multiplexing

Another method by which reconstruction problems originating from wavelength discrepancies and the reduction of speckle noise can simultaneously be achieved is holographic multiplexing by which B- or C-mode sound images are assembled in a single 3-D image volume.

Historically the first holographic multiplexing method was simply multiple exposure. Hologram from properly located transparencies made from B- and C-mode sound images were recorded one after the other on the same photographic plate. Such a multiplexed hologram after development produces all N images simultaneously, and each image appears at its proper location in the three-dimensional space. However, since only $1/N$ of the dynamic range of the recording medium can be devoted to any one exposure, multiple exposure holograms put a fraction proportional to $1/N^2$ of the available power into each image, N being the number of the images. The noise background is the same independent of N, so the signal-to-noise ratio of any image is $1/N^2$, equal to the signal-to-noise ratio achievable by a similar

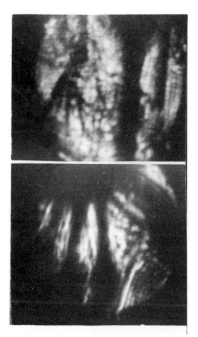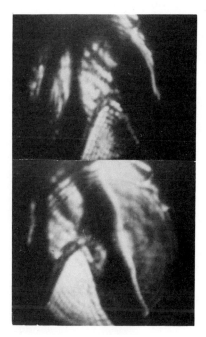

Plate 27 Sound images reconstructed from ultrasonic holograms recorded by pulsed sound and pulsed laser illumination at the Shanghai Institute of Cell Biology of Academia Sinica. The right side pictures are different views from a palm, the left side ones from an upper arm.

133

Plate 28 Reconstruction of holographically multiplexed ophthalmic B-mode images taken from another viewing angle than the B-mode images have been recorded.

hologram with $N = 1$. Since at least 10 images are needed to have a good image depth, the reconstructed image volume would be only 1/100 times as bright as the image that could have been produced by a hologram of only one B scan. Further, the reconstructed images are so inextricably linked that it is impossible to eliminate one or more of them while retaining the others. Such an operator control, however, would prove very useful.

The luminosity of the reconstructed images can be saved if the photographic plate is divided into N spatially discrete areas, each area carrying one and only one hologram of a B scan. The trouble is, however, that these areas do not coincide, so that the viewer does not see the images as simultaneously present in the same volume

Plate 29 Reconstruction of holographically multiplexed ophthalmic B-mode images taken from another viewing angle than the B-mode images have been recorded.

134

of space. The three-dimensional impression of the insonified scene can be constructed in time by projecting the reconstructions sequentially on a moving screen. However, this is a very complicated approach which requires elaborated equipment, so it is unlikely to have practical value.

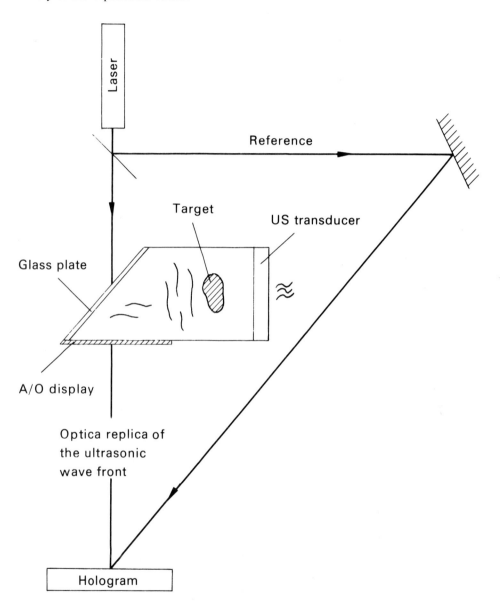

Fig. 59 Schematics of an arrangement to record light holograms of visualized sound images.

Both Greguss and Caulfield[214, 215, 216], arrived independently and practically at the same time at the idea of sharing the full aperture of the photographic plate among all the wavefronts by placing spatially complementary pseudo-random sampling masks immediately adjacent to the recording medium during recording. Each mask has $1/N$ of its area transparent and $(N-1)/N$ of its area opaque. The open areas of the mask are chosen according to the following criteria:

a) every area on any one mask must not be an open area on any other mask,

b) every area must be open on one of the masks,

c) the point spread function of each mask must be narrow with low smooth side lobes, and

d) there must be several open areas in each mask through which both eyes of the viewer in the field of view can see the image.

The size of a subdivision is set by the minimum size of hologram. Assuming somewhat arbitrarily that at least 100 fringes should be recorded and that only 1000 mm is used, then the minimum subdivision size is 0.1 mm. A practical limitation is set by the accuracy with which the mask can be made and aligned. Certainly, 1 mm is fairly easy, and 0.1 mm is quite difficult. Thus, the practical limitation lies in the order of 0.1 mm.

The reconstruction of the recorded wavefront is performed without these masks, and the resulting image is a fully three-dimensional view of the multiplexed B-mode images. The viewer focuses on the image and hence does not attend to the hologram plane where the scattered light reveals the checkerboard pattern (Plate 27). Plates 28, 29 and 30 show photographs of reconstructed images obtained from an opthalmic B-scan image volume, taken from a viewing angle other than where B-mode images have been recorded from.

5.4.3. *Light holography of visualized sound images*

The unique property of a hologram of having the capability of storing more than one independent wavefront and reconstructing them simultaneously can be used not only to assemble a 3-D image volume of spatially arranged 2-D acoustic images, but also to perceive more information from the visualized sound image. In 4.2.7 we already indicated that an optical hologram recorded from the black-and-white sound image can help to eliminate the background noise in the visualized sound image resulting from the unshifted part of the reconstructing light.

Recently it was suggested that using light holography to record visualized sound fields the frequency dependence of sound scattering properties of the insonified target could be studied[217]. The schematics of such a simple arrangement is shown in Fig. 59. A part of the laser beam used to visualize the scattered ultrasonic field is used as a reference beam by inserting a beam splitter B and a reflecting surface mirror M into the illuminating beam before it reaches the area detector so that a hologram of the visualized sound field can be recorded on the photographic plate. Two exposures,

136

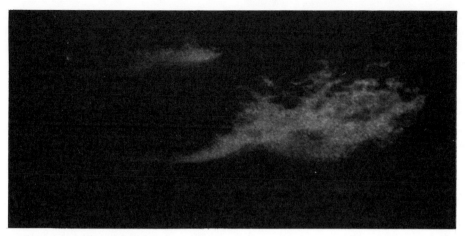

once at insonifying frequency N_1 and once at insonifying frequency N_2 will yield a double exposure hologram which, when reconstructed, will show—similar to double exposure hologram in holographic interferometry—any changes in the scattering pattern. From this the conclusion can be drawn concerning the scattering structure of the specimen.

The drawback of this technique is only that the hologram plate does not 'remember' which exposure was first, which makes it difficult, if not impossible, to evaluate the

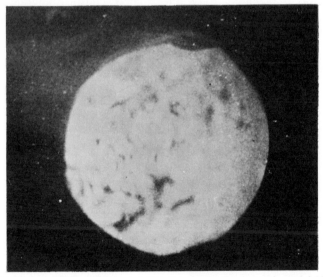

137

reconstructed image. This problem can be eliminated by using the so-called sandwich technique[218, 219]. In sandwich hologram interferometry two plates are used in one plate holder or two plates of different speed are glued together. During the first exposure the back plate is covered with a slip of black paper introduced into the gap between the plates, and so at this time only the slow emulsion facing the object is exposed. The second exposure is of much shorter duration and results in a hologram recording on the back plate without appreciably affecting the front plate. Plate 31 is the photography of a reconstructed sandwich hologram from two scattered sound fields with slightly different frequencies.

5.4.4 Sound reconstructed acoustic holography

All problems arising from wavelength discrepancies could be eliminated if acoustic waves having the same wavelength as those which were used to record the hologram were used for reconstruction, but then the problem of visualization of the recon-structed acoustic image presents itself in a new form. And even if solved, it will be only two-dimensional. There are, however, situations where the information content of the acoustic holograms can be utilized without visualization of the reconstructed acoustic image, for instance, if the acoustic hologram is thought to serve as a complex filter in a pattern recognition task, since the relation between a complex filter and a Fourier transform hologram is well known.

As demonstrated in the previous chapters, several established techniques exist to record and visualize the interference pattern of an acoustic hologram, however, there is a lack of appropriate technique to produce a good acoustic hologram for operating directly in an acoustic field. The acoustic holograms recorded by the conventional methods do not diffract ultrasonic waves because the thickness of these recording

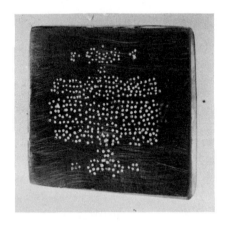

Plate 32   A binary mask for filtering in ultrasonic domain.

138

elements is not in the order of the acoustic waves which are thought to act as the reconstructing wavefront.

To overcome this problem Greguss[220], proposed to transform the recorded and visualized ultrasonic hologram by an etching and/or engraving technique well known in typography, or to generate binary holograms and binary masks with the aid of computers and plotters. Plate 32 is the photograph of such a mask. The relationship of the size and spacing of the small holes in an ultrasonic absorbing material with respect to each other contains the amplitude and phase information. This idea was then developed further by Pfeifer[221], for rapid non-destructive testing of assembly line parts for internal flaws, misdrilled holes, etc. The quality of these acoustic hologram masks allowed, however, only an accept-or-reject decision, but if further developed, it could be used for more differential testing.

Perhaps the unique technology suggested by Mailer *et al* could yield a good acoustic hologram for acoustic reconstruction. According to this idea solid particles are suspended in a medium which is solid at room temperature, and liquid above 85°C or so. The formation of the acoustic hologram then takes place at a temperature above the melting point of the medium, according to a similar mechanism in a Pohlman cell. As the medium solidifies by lowering the temperature to room temperature, the particles become trapped in the particular position which they had occupied due to the acoustic interference field.

This idea could perhaps be further pursued by omitting the suspended particles and using a thick thermoplastic material. If it is heated until it is soft, it will be deformed according to the ultrasonic interference pattern, and when cooled the spatial deformation will be frozen into the material. Most probably the acoustic intensity required to form such a phase hologram for acoustic reconstruction will be rather high, nevertheless, probably not so high as to be worthless to exploit this idea which may become important if data processing in the acoustic domain is taken into consideration.

5.4.5 *Computer-aided techniques*

Computers can be used in a variance of forms to display acoustic images in a visual form, but they all have in common that during this procedure a scanning operation is performed and a detector is quantized by an A/D converter so that nonmonotonic and irreversible nonlinear distortion of the detected data is introduced[222]. The previously described methods of analyzing the effects of scanning on image quality can also be applied to quantization nonlinearities. According to those sampling laws, to avoid aliasing errors it is necessary to sample at intervals d which satisfy

$$d < 1/2 \ N_{max} \qquad\qquad\qquad\qquad\qquad\qquad\qquad \textbf{72}$$

where N_{max} is a maximum spatial frequency contained in the intensity pattern incident on the detector. However, Equation *72* is not a sufficient condition to guarantee freedom from aliasing effects. Quantization may introduce high frequency components. Knowing in advance the characteristics of the detector, if they are monotonic, one can compensate for the nonlinearities by appropriate modification of the quantization level on the A/D converter but for those inherent in quantization cannot, because they are irreversible. This has to be kept in mind if one or another computer-aided technique is considered for visualizing sound images.

5.4.5.1 Computer-aided acoustic hologram reconstruction. In acoustic holography the display of three-dimensional virtual image of the insonified target is not feasible in general, only a tomographic display of the different sections is possible. Therefore the use of a computer to reconstruct images of acoustic holograms on a CRT screen must seriously be considered, since in this case the problems of wavelength discrepancy are nonexistent. Goodman[223], was the first to demonstrate the feasibility of this concept. The technique itself is rather simple. The acoustic hologram is presented to the computer input either in the form of digitized electric signals issuing from the scanning acoustic sensor or from the photosensitive surface of a vidicon onto which the transparency of the visualized acoustic hologram pattern has been imaged. The main limiting factor in the reconstruction of this acoustic hologram is the computer time needed to perform the digital Fourier transform operations, especially if 'focusing' on different image planes is desired. So, for example, the computation time for a two-dimensional array of about 10 000 elements is about 15 minutes, which is an acceptable time in several practical applications, such as underwater viewing. As computer technology advances, the time will undoubtedly be reduced and digital image reconstruction will be used in general, especially in medical diagnostics.

5.4.5.2 Ultrasonic transmission tomography by reconstruction Sound images formed by transmission are similar to X-ray images in the sense that they convey information on the apparent average sound absorption property of the target (3.6.2). Such an information could be especially important in medical diagnostics since there is a large difference in ultrasound attenuation between various soft tissues. Nevertheless, up till now, most attempts have failed to use this imaging technique due to practical difficulties, including scattering and phase distortion by bones, air spaces, etc.

There is, however, a possibility of overcoming these difficulties if scanned transmission images of a transverse plane of the body are formed from a large number of angles with subsequent reconstruction of the transmission image by a computer, similar to X-ray computerized tomography.

The theorem that is fundamental to understanding the reconstruction from projections is called 'central-slice' or 'projection-slice' theorem. It has wide utility from radio astronomy to seismic imaging[224]. Greenleaf *et al*[225] applied it first to ultrasound and succeeded in imaging soft tissue samples and resolution test objects which did not obstruct transmission of ultrasound beam as bones in the body do. Carson *et al*[226, 227], improved this technique by developing both an appropriate instrumentation and also a good reconstruction algorithm.

When an ultrasonic pencil beam of intensity I_o passes through a homogeneous medium, it is attenuated according to Equation **30**. For nonhomogeneous objects the exponent is replaced with the line integral of the linear absorption coefficient $\alpha(x,y)$.

$$I = I_0 \, \exp\left[\int_{source}^{detector} \alpha(x,y) \, dl\right] \qquad\qquad 73$$

To linearize algebra the negative logarithms of the raw data have to be taken which then gives the integral equations of the form

$$f_\phi(x') = -\ln(I/I_0) = \int_{source}^{detector} \alpha(x,y) \, dy' \qquad\qquad 74$$

If pulsed ultrasound is used, then we have another measurable variable, pulse transmit time ($t - t_o$), which has a simple dependence upon the velocity of ultrasound (c) and the distance (x) travelled through the medium

$$t - t_o = x/c \qquad\qquad 75$$

In the first case ultrasound transmission tomography of attenuation by reconstruction (UTTAR), in the second case ultrasonic transmission tomography of velocity reconstruction (UTTVR) is formed.

The problem is to obtain an estimate of $\alpha(x,y)$ designated $\bar\alpha(x,y)$, [or $c(x,y)$ designated $\bar c(x,y)$] from a sufficiently large set of projection data $f_\phi(x)$. To solve it each transmission scan which is a one-dimensional intensity or velocity profile, a projection, must cover the entire cross-section of the body in the plane imaged. Scans of the plane must be obtained at numerous angles about the line perpendicular to the plane. The one-dimensional Fourier transform of a one-dimensional projection of a two-dimensional object is, however, mathematically identical to a slice (line) through the two-dimensional Fourier transform of the object itself. Therefore, from all one-dimensional projections the two-dimensional transform of the object can be synthesized and from this the object is readily obtained by an inverse two-dimensional transform. However, this is not the only possibility of reconstructing the sound image from the above data. The algorithms capable of producing satisfactory reconstructions are reviewed by Gordon and Harman[228].

The capability of this technique is shown in Plate 33. This is a photograph of an ultrasound transmission tomographic image of the attenuation coefficient through a plane in the breast of a 24-year-old woman taken by Carson[229]. It was obtained by 3.5 MHz, 10 mm diameter, 9 cm focal length transducers moved on either side of the breast in a rotate-translate motion using equipment described in [226] and [227]. The very dense subcutaneous connective tissues are revealed as the thick bright line in the upper portion of the breast and on the sides. These connective tissues have an apparent attenuation of up to 5 dB/cm/MHz. In the lower right is a circular area with

141

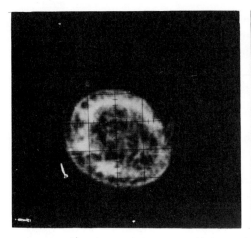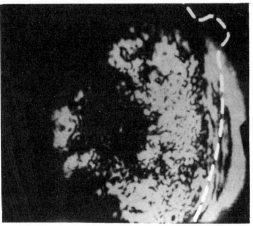

Plate 33 Ultrasonic transmission tomographic image of attenuation by reconstruction of a woman's breast recorded in vivo[229] and the reconstructed acoustic holographic image of an excised breast, recorded with surface levitation technique (*courtesy* of Holosonics, Inc.)

stripes radiating from its centre, suggestive of normal glandular architecture. Comparing this acoustic image recorded *in vivo* with the acoustic holographic image of an excised breast, recorded with surface levitation methods, we think that there is no doubt which of the two techniques has a more promising future for imaging soft tissues such as a female breast. We are sure that UTTR will also provide significant new information in body planes containing bones. Naturally, there remain *unanswered* questions concerning UTTR—it is not clear what difference in soft tissue attenuation coefficient will be detectable in a plane containing bones, due to uncontrollable scattering and the reflection situation.

5.4.5.3 Computer sectioning A computer technique has recently been developed by Robinson[230], which enables the reconstruction of usable B-mode images in a plane perpendicular to a set of transverse B-mode images, if they had previously been digitized and stored in the memory of a computer. By combining this derived plane with information obtained by an actual scan in this plane, further information can be added into the display available to the interpreter, thus the loss of information due to specular reflection can be overcome to some extent. By multiple angular sectioning B-mode images from any plane can be obtained.

The generation of these B-mode images is quite simple. First, the original B scans have to be digitized. Since to date gray scale B-mode images contain about 1024 x 1024 points and 8 bit gray scale, the amount of data has to be reduced to about 100 x 80 points and 6 bit gray scale. The pictures are then stored in one-dimensional arrays to allow a greater flexibility in picture shape manipulation. Care has to be taken that the data points on one picture correspond with those on the other pictures after digitalization.

To avoid 'dotty' appearance of the displayed image every picture point is displayed four times to form a four-dot square. The final registered picture contains in general

142

100 x 130 points to allow the overlap that is necessary for correct registration. To generate a section of right angles to the input plane, a line of points on the intersection of each input plane and the section plane is taken. If the number of the input B-mode images is n, the new section will consist of n lines, spaced 5 picture points apart, with approximately 100 points in each line, depending on the angle of the section.

Obviously, these procedures lead to resolution degradation. But if we consider that at right angles to the ultrasonic beam the resolution is in the order of a few centimeters, and that a B-mode image represents in reality not a single section, but a slice of 1–3 cm thickness, this degraded resolution may well be sufficient for a preliminary interpretation to determine the areas of interest where full resolution is needed. In contrast to other computer techniques, the time required is in the order of 1–3 sec per section and not minutes. Neverthless, it is unlikely that this method would find application as long as mechanical scanners are mainly used to form B-mode images, but when sound imaging techniques are considered as described, e.g. in 5.3, this computer technique of displaying three-dimensional ultrasonic data may lead to a new generation of 'seeing-by-sound' equipment.

5.4.6 3–D Impression via 'linearity' and 'isochronicity'

The new direct ultrasonic image display method introduced recently by Hanstead[231], is based on the acoustic analog of a little known optical phenomenon, whereby a coaxial assembly of two identical converging lenses can cause a 3–D object to produce a 3–D image of the same shape[232]. As seen from Fig. 60 a spatial array of luminescent points will show up in an identical array of images, but with transverse inversion. It can be seen only from the system's exit pupil, *and only from there*, because each point image is formed from a cone of rays which converges on approaching the image, and diverges on leaving it. This limitation, however, disappears if acoustic focusing is used, provided

a) the acoustic images display the same spatial relationships as the original object; the property is called here 'linearity',

b) the acoustic transit time from transducer to object to image is the same for all objects in the three-dimensional field; the property is called here 'isochronicity'.

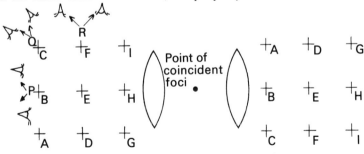

Fig. 60 Schematics of the optical phenomenon called 'linearity'.

143

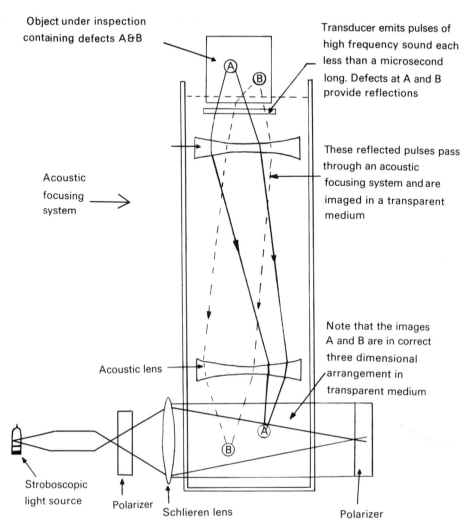

Object under inspection
containing defects A&B

Transducer emits pulses of
high frequency sound each
less than a microsecond
long. Defects at A and B
provide reflections

These reflected pulses pass
through an acoustic
focusing system and are
imaged in a transparent
medium

Acoustic
focusing
system

Note that the images
A and B are in correct
three dimensional
arrangement in
transparent medium

Acoustic lens

Stroboscopic
light source

Polarizer

Schlieren lens

Polarizer

Fig. 61 Schematic diagram of direct 3-D visualization of a sound field.

These necessary, but sufficient conditions can be achieved if

a) the nearer focal points of the two acoustic lenses are coincident,

b) the farther focal lengths are in the ratio of $1:\sqrt{2}$, the larger one being that nearest the image.

To render the so-called 3–D sound image visible, several visualization techniques can be used, such as schlieren, photoelasticity, etc., and binocular vision will reveal any three-dimensional information present in the image field. The basic layout of such a system is shown in Fig. 61. The insonification is provided by short pulses of ultrasound generated by the transducer, any reflections from the insonified space

144

pass back through the transducer and are focused in the image space. A transparent imaging medium is used, and by placing crossed-polarizing filters on either side of it, the stresses associated with ultrasound become visible as bright areas against a dark background if it was a solid material.

The light source used is stroboscopic and synchronized to the ultrasonic pulses with suitable delay interposed. Thus, if the light flashes are short enough to freeze the advancing ultrasonic wavefronts, the eye will see the discontinuities correctly imaged, displaying the original spatial relationships. Binocular viewing will reveal any three-dimensional information present in the image field. The great advantage of this method is that it does not need coherent illumination and that real-time 3–D continuous imaging is possible.

The technique has already been used successfully in nondestructive testing. More details for the practical design of a system can be found[233].

5.4.7   3–D Impression via optical tricks

The efforts to display simultaneously altitude, azimuth and depth (range) even if it is only an illusion of 3–D have not been given up when it turned out that acoustic holography would never produce a 3–D image because of the differences in wavelengths between the insonifying beam and the laser light used in reconstruction. Several techniques applied in the design of binocular instruments and various optical tricks have been tried and are still investigated to get something comparable to a holographic display concerning three-dimensionality.

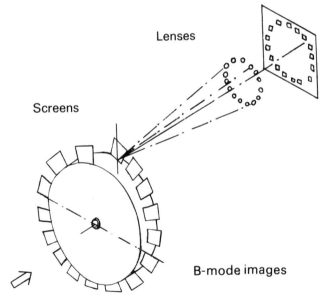

Fig. 62   Schematic of a 3-D display system according to Szilard.

145

One approach has been suggested by Szilard[234]. As seen in Fig. 62, a number of plane ground glass screens are mounted around a helix in planes perpendicular to the axis of the helix around the surface of a disc. When this disc is rotated, the screens will pass one after the other through the same angular position but in different 'depths' in relation to the viewer, who looks in the direction of the axis of the helix. The B-mode sound images are projected on the screen of a corresponding 'depth' while it passes through the appropriate position, and since the disc is rotated with a sufficiently high frame rate, three-dimensional image volume of the insonified space will be perceived.

Since the B-mode images represent a section as thick as the diameter of the insonifying beam, true three-dimensional impression will be achieved only if the screens are of the same thickness. The advantage of this technique is that up to 75° off-axis views are possible, and that the brightness of the layers at different depths can be varied independently of one another. The disadvantage is, however, that it is not a real-time procedure, and the system is rather bulky.

Another version of this idea, Groh[235], has been applied up till now only to X-ray images, but there is no reason why it could not be used to display a three-dimentional image volume composed of B-mode sound images. The principle of this method is shown in Fig. 63. Not the B-mode sound images, but their optical holograms are circularly arranged. The recording of these holograms is performed by projecting them with a laser beam on a translucent diffusing screen, and one part of the projecting laser beam is used as reference background to form the hologram.

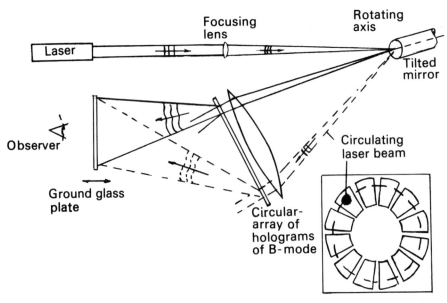

Fig. 63 Holographic multiplexing using circular synthetic aperture.

146

On reconstruction with a reference wave the three-dimensional virtual image of the insonified scene will be seen in front of a brightly illuminated screen, and it will be orthoscopic. The depth of details of the object can be estimated with about the same accuracy as in ordinary holography, and more precisely by utilizing the real images used for the reconstruction of the conjugate complex reference wave. The depth can then be scanned just by moving the projection screen.

The method has the further advantage of reducing the annoying speckle noise by simply employing a rotating tilted mirror for generating the reference beam in the reconstruction process.

Although the method is very elegant, it suffers from the same disadvantage as the previous one, but it has the merit of delivering a permanent three-dimensional record from the insonified target.

To assemble in real time a three-dimensional image volume from B-mode sound images a varifocal mirror system has been proposed by Greguss[236]. The varifocal mirror is an aluminized mylar film stretched over a loudspeaker and it continuously changes its focal length if a sinusoidal voltage is applied to the loudspeaker. The B-mode images displayed for instance on a CRT screen are then projected, one after the other, through a beam splitter on this surface, as illustrated in Fig. 64. If the mirror pole displacement $\triangle$ is much smaller than the CRT-mirror distance as shown in Fig. 65, image distance or the mirror diameter, the most important relationship between the imaging parameters can be described according to Rawson[237], by the following equations:

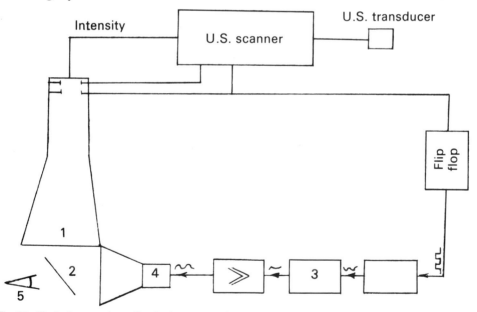

Fig. 64 Block diagram of a varifocal mirror system for B-mode image display. 1. CRT tube. 2. Semitransparent mirror. 3. Phase shifter. 4. Loudspeaker with mylar. 5. Observer.

$$\sigma' = -\sigma/(1 + 4\sigma\delta)$$
$$m = 1/(1 + 4\sigma\delta)$$

76

where

$$\sigma = \frac{S_F}{R} \;\; ; \;\; \sigma' = \frac{S'_F}{R} \;\; ; \;\; \delta = \frac{\Delta}{R} \;\; ; \;\; m = -\frac{S'}{S}$$

since the spherical mirror equation

$$\frac{1}{S} + \frac{1}{S'} = \frac{2}{r}$$

77

holds only where S and S' are the distances of the CRT and image, respectively, measured from the center of the display's diaphragm, r is the radius of curvature of the mirror, and R is the radius of the mirror.

It is selfevident that as the image moves along the depth axis toward the observer, the image diminishes in accordance with the magnification $m = S'/S$. Thus, for the reflected image to be of constant dimension the size of the two-dimensional display on the CRT must be inversely proportional to the instantaneous magnification, which can be assured by appropriate electronics. Since the mirror vibrations are symmetrical to O_f, m oscillates about unity. We have to black-out one half of the effect of the focal point shift. This means that having displayed n B-mode images in t seconds, we have to black-out the screen for the same time duration. It will allow the image plane

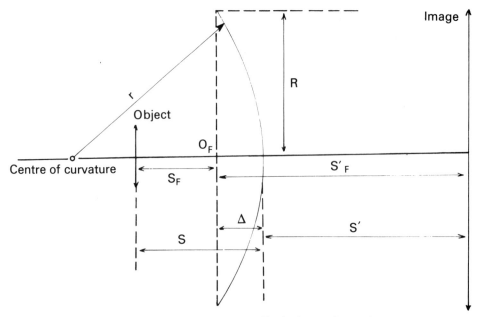

Fig. 65 Principle of the vibrating varifocal mirror performance.

148

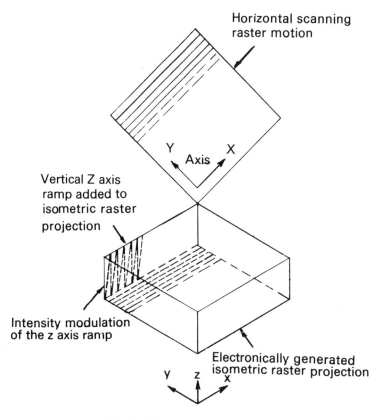

Horizontal scanning
raster motion

Y
Axis
X

Vertical Z axis
ramp added to
isometric raster
projection

Intensity modulation
of the z axis ramp

Electronically generated
isometric raster projection

y   z
x

Fig. 66   Schematics of isometric imaging.

to retreat to its starting stage. Only in this case is a three-dimensional image volume assembled from the spatially distributed sequence of B-mode sound images.

Since the required sequence of displaying the B-mode images is rather high, mechanically scanned systems are difficult to combine with a varifocal mirror technique. However, this method may have a great future in assembling B-mode sound images in a 3–D image volume which were produced by one or other electronic scanning technique.

5.4.8   Isometric imaging

Isometric imaging of an insonified target, proposed by Becker and Trantow[238], is making the illusion of three-dimensionality on a two-dimensional surface in a somewhat similar way as an artist represents space. According to their suggestion, the 3–D image is displayed on a two-dimensional screen of a memory scope by adding a ramp signal to the voltages describing the x and y position of the scanning transducer and

149

by modulating the intensity of the electron beam. The depth then corresponds to a particular height of the z axis ramp as shown in Fig. 66. The x, y and z coordinates of the scanned sound space are converted to x and y of the display scope according to Equations **78** and **79**:

$$x' = y \cos\Theta + y \sin\Theta \qquad\qquad\qquad \textbf{78}$$

$$y' = (y \cos\Theta - x \sin\Theta) \sin\varphi + 2 \cos\varphi \qquad\qquad\qquad \textbf{79}$$

where Θ is the rotation of axis x,y, and φ is the tilt of axis z. If Θ and φ equal zero, a B-mode image is displayed, and if Θ equals zero but $\varphi = \pi/2$, it will be a C-mode image.

To eliminate unwanted signals time-gating is used. If x, y and z information is recorded on an endless magnetic tape loop and then played back at high speeds, the insonified target can be reviewed from several perspectives without rescanning the target.

The advantage of this display method is that it has the ability of shifting the perspective, and that the depth information is in focus through the entire field of view. Its disadvantage is the disadvantage of every C-scan method – it basically *is* a C-scan technique – and that is, its time consuming value. But perhaps by adapting this idea to methods described in Chapter 6.4, a new versatile generation of seeing-by-sound device can be developed.

5.4.9   Color displays

The general approach in displaying visualized sound images is to discover and employ selectively inherent capabilities of the human visual system in such a manner as to improve the observer's performance in detection, recognition and discrimination of the details in the display. This is one of the reasons why quantized[239], and gray scale B-mode images[240], have become more and more popular, especially in medical diagnostics. It is said, however, that the human visual system is able to distinguish simultaneously only about 15 shades of gray from black to white, thus much recorded information in an image may be lost to the human observer. It is also often said that for an equivalent image in full color, the human observer can readily make many times this number of discrimination. So it is not surprising that the efforts for color display of visualized sound images are perhaps as old as the endeavors for creating an efficient and economical acoustical-to-optical display, even if this 'colorfulness' means only a better intensity discrimination, and not wavelength discrimination, as in optics.

The statement that color changes can be better interpreted than brightness changes is, however, subject to considerable misinterpretation, which is the result of misunderstanding of the psychophysics of color perception.

First of all, although it is true that the range of luminance levels to which the

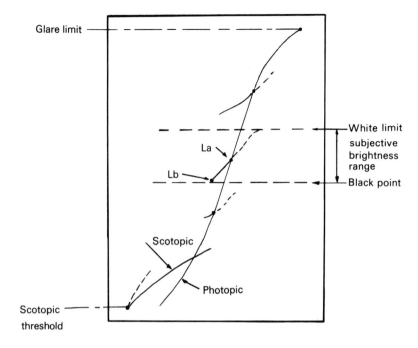

Fig. 67 Adaptation range and subjective brightness range for human vision.

human visual system can adapt is in the order of 10^{10} from scotopic threshold to glare limit, however, it cannot operate over such a range *simultaneously*. The total range of luminous levels it can distinguish simultaneously is rather small. There is always a luminous level below which all stimuli are perceived as indistinguishable black as shown in Fig. 67. The brightness area called white limit cannot be extended without moving the mean adaptation level to a new higher value. The gray steps in the displayed image do not correspond to the distinguishable luminance depth obtained under conditions of complete adaptation at each luminance level, but rather it is adapted to a mean luminance level. As the eyes rove about the displayed image, the instantaneous adaptation level fluctuates about this mean. In analogous fashion, this holds for color perception too, but the situation is still more complex. In reality, the visual system can distinguish more levels of information in color than using the gray scale alone, because it uses simultaneously brightness and chromatic variations. The evaluation of the simultaneous brightness and chromatic variations may lead, however, to misinterpretation if too much importance is attributed to the perceived color. So, if a mixture of red and green is displayed and it is viewed at high brightness level, the sensation of yellow is perceived. However, the very same mixture of red and green viewed at low brightness level will produce the sensation of brown.

Thus, there are two distinct problems in designing color displays for visualized sound images:

151

a) The assignment of a particular chromaticity and luminance to a particular 'sonic gray scale' is the subject of psychophysical research as to the effects of combined color brightness contrast on image discrimination tasks.

b) Once the assignment has been made, either by personal choice or from the result of psychophysical research, the appropriate techniques have to be found to produce the specified chromaticities and luminance values.

Concerning problem (a), practically no research, that is no systematic research has been performed as yet. It was suggested to use the 'color stereo' effect to eliminate at least partially the first order distortion in the reconstruction of acoustic holograms[241]. It is based on the observation that if the outer part of the eye pupils are covered, blue images seem closer than red images, and images in other colors seem to lie somewhere between the two, according to their wavelengths[242]. The reason for this strange experience is that the lenses act if they were prisms with bases facing outwards, and so blues are deflected outward more than other colors. As a consequence the eyes must converge more to bring blue in focus than to focus red images, and a stereo feeling develops.

It was also proposed to apply the results of color image enhancement used in aerial photography[243], but its adaptation to sound images requires caution.

Concerning problem (b), basically two directions can be followed:

1) the B-mode image is in the form of a black-and-white or gray scale transparency, or

2) the electric signals of the sampling aperture are directly fed to the input of the color CRT.

5.4.9.1 *Sound image transparencies*   Sound image transparency for pseudocolor transformation can be formed either by one of the techniques described in 4.1 or by photographing the CRT screen. Independently from the way the transparencies have been produced, the first step is to determine whether the gray scale to be transformed into color is corresponding directly or inversely to the intensity of the insonifying beam. Although the ratios of the intensities in the sound image are unchanged, the relative intensity distribution in the transparency can be modified extensively by the choice of the film material, exposure time, developing technique, etc. So the transparency may show a higher level of gray scale than there is in reality in the absorption properties of the insonified object, or vice versa. This may lead to misinterpretations, especially when transformed into color.

Technically this transformation can be performed in several ways. One of these is the pseudocolor three separation process of Stratton and Sheppard[244]. This is a relatively inexpensive but rather time-consuming method of producing pseudocolor transformation by a photographic procedure.

Another possibility to convert the gray scale of the transparency into color is the digital approach. It has the advantage of being extremely flexible and of being rather troublefree and easy to use once the programming of the system has been worked out[245]. Its disadvantage is that the source of noise includes not only the noise of the transparency but also that of the scanning digitizing system. While in the first

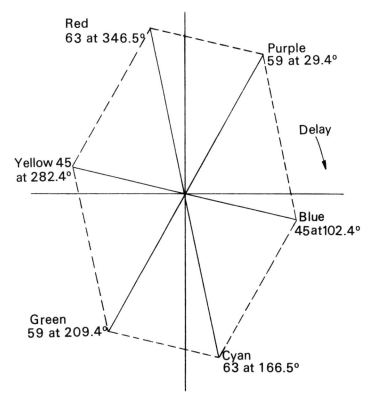

Red
63 at 346.5°

Purple
59 at 29.4°

Delay

Yellow 45
at 282.4°

Blue
45 at 102.4°

Green
59 at 209.4°

Cyan
63 at 166.5°

Fig. 68 Vector diagram showing ultrasound carrier amplitudes and phase corresponding to six major colors[246].

method there is no ambiguity in the designation of the particular chromaticity and luminance level, here it is the signal-plus-noise which is quantized, so there is some uncertainty in the choice of chromaticity and luminance value.

5.4.9.2 *Color CRT displays* Jacobs *et al*[246] were perhaps the first who used color-enhance technique to improve the perception of acoustic impedance changes in visualized sound image displayed on a CRT screen. The signals of a Sokolov type ultrasonic vidicon camera were adapted to the NTSC system of video signal processing, the standard for color television within the United States. The output of the ultrasonic camera has been displayed as a varying amplitude signal whose color or hue is indicative of the phase delay encountered by the signal at the particular point of the scanned ultrasound image. As illustrated in Fig. 68, a signal having a given amplitude will, depending upon its relative delay in reference to a burst signal, produce a color that is defined by this phase angle with an amplitude or level that corresponds to the signal strength.

It is claimed that the sensitivity of this system is so high that changes in acoustic impedance may reliably be detected in the order of one part in 10^8. This means that

153

changes in acoustic properties due to an increase in free electron population of approximately $10^{13}/cm^3$ will show up in color change so that the system can be used to investigate the free electron distribution over extended areas of semiconducting layers.

Basically the same idea, without phase reference, has been adapted to mechanically scanned ultrasonography too, using for instance a minicomputer to memorize the digital signals of the A/D converter, and then, if required, the sound image can be displayed in color[247]. There is, however, the question whether this sophisticated technique gives really more information or is only a commercial gimmick.

References

[171]SNYDER, H. L. (1967) *USAF Report* AFAL-TR-67–293

[172]POOLE, H. H. (1966) *Fundamentals of Display Systems* Spartan Books, New York.

[173]LUXENBERG, H. R., KUEHN, R. L. (1968) *Display System Engineering* McGraw-Hill, New York

[174]SZEBENI, A., SINKOVICS, I., TARJAN, GY. (1978) II. Congress of the Hungarian Society for Nuclear Medicine, Budapest

[175]JOHNSON, J. (1958) Image Intensifier Symposium, Fort Belvoir, Va.

[176]LAPORTE, H. R., CALHOUM, R. L. (1966) *Autonetics Technical Report T6–2258/501*

[177]DAVIDSON, M. (1968) *J. Opt. Soc. Am.* **58** 1300–1303

[178]THOMPSON, F. T. (1957) *J. SMPTE* **66** 602–606

[179]KRAUTKRÄMER, J. (1959) *Arch. Eisenhüttenw.* **30** 693–703

[180]GERICKE, O. R. (1963) *J. Acoust. Soc. Am.* **35** 364–367

[181]U.S. Pat 3,662,589 (1971) L. Adler, H. L. Whaley

[182]BAUM, G. (1969) *Ophthalmic Ultrasound* ed. K. Gitter *et al.* pp. 231–236 The C. W. Mosby Co., St. Louis

[183]LECHNER, B. J. (1971) *Proc. IEEE* **59** 1566–1579

[184]JACKSON, R. W., JOHNSON, K. E. (1971) *IEEE Trans. of Electro Devices* **ED–18** 316–322

[185]U.S. Pat. 3,479,946 (1969) S. C. Requa

[186]BRODY, T. P., LUO, F. C., SZEPESI, Z. P., DAVIES, D. H. (1975) *Trans. IEEE* **ED–22** 739–748

[187]GREGUSS, P. (1971) *SPIE Seminar Proceedings,* **24** 55–83

[188]BAUM, G., GREENWOOD, J. (1961) *N. Y. State Med. J.* **61** 4149–4157

[189]GABOR, D. (1951) *Proc. Phys. Soc.* **B64** 449–452

[190]CAULFIELD, H. J., SUN LU (1970) *The Application of Holography* Wiley Interscience, New York

[191]COLLIER, R. J., BURCKHARDT, C. B., LIN, L. H. (1971) *Optical Holography* Academic Press, New York

[192]VIENOT, C. H., BULABOIS, J., PASTEUR, J. eds. *Applications of Holography* Université de Besancon (1970)

[193]THOMPSON, B. J., DEVELIS, J. B. eds *Developments in Holography* SPIE Seminar Proceedings **25** (1971)

[194]RUBY, S., BHUTA, P. G. (1972) *Engineering Applications of Holography* SPIE Los Angeles

[195]ROBERTSON, E. R. (1975) *Engineering Uses of Coherent Optics* Cambridge University Press, Cambridge, London, New York

[196]WINTER, D. C. (1972) *Acoustical Holography* Vol. 4 ed. G. Wade pp. 635–640 Plenum Press, New York

[197]GREGUSS, P., BUGDAHL, V. (1973) *Jenaer Rundschau* **18** 298–301

[198]WEISS, L, HOLYOKE, E. D. (1969) *Surgery, Gynecology & Obstetrics* **128** 953–962

[199]LUCEY, G. K. Jr. (1968) *J. Acoust. Soc. Am* **43** 1324–1327

[200]NIGAM, A. K., FRENCH, J. C. (1974) *Acoustical Holography* Vol. 5 ed. P. S. Green pp. 685–700 Plenum Press, New York

[201]Fr. Pat. 3,829,827 (1974) J. Ernvein

[202]VARSHAVSKY, Y. I., BRAGINSKAYA, F. I., KRUGLYAKOVA, K. E. DADASHEV, R. S. (1974) *Doklady Akad Nauk* **217** 719–721

[203]VARSHAVSKY, Y. I., BRAGINSKAYA, F. I., BUNTO, T. V., KRUGLYAKOVA, K. E., DADASHEV, R. S. (1974) *Biophysics USSR* **19** 375–377

[204]BRENDEN, B. B. (1975) *J. Acoust. Soc. Am.* **58** 951–955

[205]BRENDEN, B. B. (1975) *Sci. Instr.* **8** 885–894

[206]ERIKSON, K. R., FRY, F., JANES, J. P. (1974) *IEEE Trans. Sonics & Ultrasonics* **21** 144–170

[207]BERBEKAR, G. (1973) *Digital Holography* Research Report, Technical University Budapest

[208]METHERELL, A. F. (1969) *Acoustical Holography* Vol. 1 ed A. F. Metherell pp. 203–221 Plenum Press, New York

[209]BERBEKAR, G., TÖKÉS, S. B. *Ultrasonics* **16** 251–258

[210]FARHAT, N. H. (1975) *IEEE PROC. Dec.*
[211]HIDAKA, T. (1975) *J. Appl. Phys.* **46** 786–790
[212]LANGLOIS, G. (1977) *Priv. Comm.* 28 Sep
[213]U.S. Pat. 3,745,814 (1973) D. Gabor
[214]REDMAN, J., WALTON, J. P., FLEMING, J. E., HALL, A. M. (1969) *Ultrasonics* **7** 26–29
[215]GREGUSS, P., CAULFIELD, H. J. (1972) *Science* **177** 422–424
[216]FALUS, M., CAULFIELD, H. J., GREGUSS, P. (1974) *Laser & Unconv. Opt. J.* No. 51 1–9
[217]GREGUSS, P. (1976) *Annals N. Y. Acad, Sci.* **267** 312–322
[218]ABRAMSON, N. (1974) *Appl. Opt.* **13** 2019–2025
[219]HARIHARAN, P., HEGEDÜS, Z. S. (1976) *Appl. Opt.* **15** 848–849
[220]GREGUSS, P. (1969) *Acoustical Holography* Vol. 1 ed. A. F. Metherell pp. 257–265 Plenum Press, New York
[221]PFEIFER, J. L. (1972) *Acoustical Holography* Vol. 4 ed. G. Wade pp. 317–322 Plenum Press, New York
[222]KAY, M., SHIMMINS, J., MANSON, G., ENGLAND, M. E. (1975) *Ultrasonics* **13** 18–20
[223]GOODMAN, J. W. (1969) *Acoustical Holography* Vol. 1 ed. A. F. Metherell pp. 173–184
[224]GORDON, R. ed. (1975) *Image Processing for 2–D and 3–D reconstruction of Projections* Stanford University, Stanford
[225]GREENLEAF, J. F., JOHNSON, S. A., SAMAYOA, W. F., DUEK, F. A., WOOD; E. H., (1975) *Acoustical Holography* Vol. 6 ed. N. Booth pp. 71–91 Plenum Press, New York
[226]CARSON, P. L., OUGHTON, T. V., HENDEE, W. R., AHUJA, A. S. (1977) *Med. Phys.* **4** 302–309
[227]DICK, D. E., BAY, H. P. CARSON, P. L. (1977) *Proc. Rocky Mountains Biomed, Eng. Symp.* 31–35
[228]GORDON, R., HARMAN, G. T. (1974) *Int. Rev. Cytal,* **38** 111–
[229]CARSON, P. L. (1977) *Priv. Comm* 31 Aug.

[230]ROBINSON, D. E. (1972) *J. Acoust. Soc. Am.* **52** 673–687
[231]HANSTEAD, P. D. (1973) *Ph.D. Thesis* The University of London, London
[232]MCLEON, J. H. (1954) *J. Opt. Soc. Am* **44** 592–597
[233]HANSTEAD, P. D. (1976) *Direct Ultrasonic Visualization of Defects* Closing Report, Central Electricity Generating Board, Bristol
[234]SZILARD, J. (1974) *Ultrasonics* **12** 273–276
[235]GROH, G. (1972) *Optical and Acoustical Holography* ed. E. Camatini pp. 100–109 Plenum Press, New York
[236]GREGUSS, P. (1972) *Proc. Symposium on Engineering Applications of Holography* eds. S. Ruby, P. G. Bhuta pp. 270–285 Los Angeles
[237]RAWSON, E. G. (1968) *Appl. Opt.* **7** 1505–1511
[238]BECKER, F. L., TRANTOW, R. L. (1972) *SPIE Proc.* **29** 61–65
[239]BAUM, G. (1972) *Ultrasonics* **10** 14–15
[240]CROCKER, E. F., MCLAUGHLIN, A. F. (1975) *Ultrasonics in Medicine* eds. E. Kazner et al. pp. 207–212 Excerpta Medica, American Elsevier, Amsterdam
[241]GREGUSS, P. (1976) *Acoustics and Vibration Progress* Vol. II. ed. R. W. B. Stephens & H. G. Leventhall pp. 1–54 Chapman & Hall Ltd., London
[242]VON RÖSCH, S. (1954) *Praxis der Physik, Chemie, Photographie* **3** 331–336
[243]GAZLEY, C., RIEBER, J. E., STRATTON, R. H. (1967) *Astro & Aero J.* **5** 56–60
[244]STRATTON, R. H., SHEPPARD, *US Air Force Project* R-596PR.
[245]BILLINGSLEY, F. C. (1970) *Appl. Opt.* **9** 289–300
[246]JACOBS, J. E., REIMANN, K., BUSS, L. (1968) *Materials Evaluation* **8** 155–166
[247]YOKOI, H., TATSUMI, T., ITO, K. (1975) *Ultrasonics* **13** 219–224

6 SEEING BY SOUND

Sound is an information carrier similar to light but it may penetrate media which are opaque to the entire visible spectrum of the electromagnetic radiation. In the preceding chapters we have seen that both information carriers are capable of forming an information pattern which is called image, only those formed by sound waves have first to be converted into a form perceivable by the human visual information processing system, and that this can be achieved by more than one way. Thus mankind now has the possibility to look into objects that are opaque or to see through water that is murky.

6.1 Under water

Underwater imaging systems offer many advantages over conventional oceanographic search and surveillance techniques such as direct optical viewing and conventional echolocation methods. Optical viewing, if not made impossible by turbid water, is limited by relatively short ranges, and echolocation does not provide detailed target information, only range information.

The idea of echoranging came to birth after the Titanic went down in 1906, and the goal was to detect icebergs in advance. Later it played an important role in defense against submarines, especially when sonar was combined with B-mode imagery, which gave a visual presentation of the distance and bearing of the submarine.

6.1.1 Fish-finding sonars

The first nonmilitary use of sonars was their use not only to detect fish schools but also to 'see' their spatial dimensions. This means that beside accurate determination of the distance between the school and the ship, complete *penetration* of the information bearing acoustic signal through the school, horizontal school width, and accurate determination of the degree of arc occupying the school is needed (Fig. 69). From these data the cross-section of the school perpendicular to the scan beam can be calculated by

$$D = R \ \mathrm{tg}\theta \qquad\qquad \textbf{80}$$

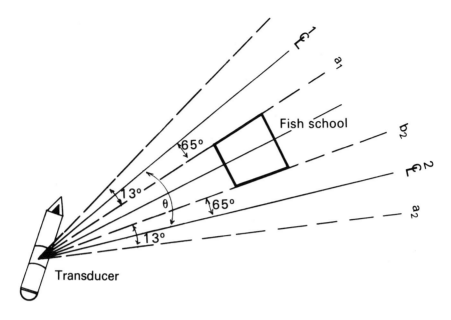

Fish school

65°

13°

θ

65°

13°

Transducer

Fig. 69 Sonar beam pattern initiating fish school echoes during scans.

where D is the diameter of the cross section, R the range from the ship to the near side of the fish school and θ is the angle between the beam position of the first and last detected echoes.

It is, however, not enough to see the fish school but is also important that the species can be identified. Theoretically this can be accomplished on the basis of their reflective coefficient, but in practice it is rather difficult because, for instance, cuttle-fish, plankton and whitebait both have similar reflective coefficient and similarly pelagic tuna and bonito fish.

Further, such factors as propagation damping characteristics of sound waves in the sea, the relative speed of the ship to the fish, etc., have to be taken into account. To accomplish this task gray scale display is essential, and even color display, if for instance tiny fish schools in plankton or around cliffs where the continental shelf bulges into the ocean have to be detected. In this case the color display indicates the same number of intensity levels as a gray scale display, nevertheless, it allows a better recognition and discrimination because the human visual system can better detect and compare isolated color patches than isolated gray scale areas[248, 249].

6.1.2 Side-scan sonars

Side-scan sonars have been developed for imaging the ocean bottom in a similar way as side-looking radars are mapping the terrain[250, 251]. The first problem of applying

157

this method for acoustic imaging arises from the difference of more than five orders of magnitude in propagation speeds between radar electromagnetic signals as compared to sonar acoustic signals. The major consequence is the difficulty of achieving an adequate unambiguous range. Since techniques for ambiguity avoidance have now been developed, it is not surprising that side-scan sonars, called also synthetic aperture sonars, start to play an important role in underwater viewing. As a consequence, an extensive literature on this topic has joined the also extensive side-looking radar literature[252–257].

The synthetic aperture method is practically a two-step method as holography is, since the acquired data have to be first stored and then processed to produce an image. The principle can well be understood from Fig. 70. For simplicity we consider continuous waves and not impulse sound, thus ignoring the problems introduced by sampling.

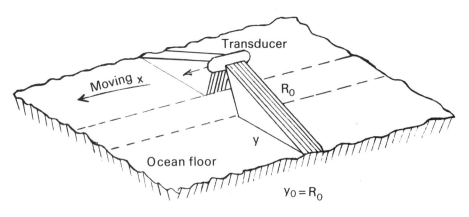

Fig. 70 Side-looking sonar geometry.

It can be shown that a signal issuing from a point reflector located on axis y at $y = R_o$ has the form at x

$$A_R = kA_o \cos\omega \, (t + 2r/c) \qquad\qquad \textbf{81}$$

if the wave transmitted from x has had the form of

$$A_T = A_o \cos\omega t \qquad\qquad \textbf{82}$$

where ω is the angular frequency, c the sound velocity in propagation medium, and k a constant.

If a coherent reference signal is added, then, after filtering, the form of the detected signal will be

$$A_D = K'A_0 \cos\left[\frac{4\pi}{\lambda}(x^2 + R_0^2)^{1/2}\right]$$

This is, however, the equation of a one-dimensional Fresnel-zone plate pattern which over a fairly large range of x is sinusoid and whose frequency is a linear function of x for any particular value of R_0. At any value of R_0 the synthetic aperture signal for a real target will consist of an array of overlapping Fresnel-zone functions. To reconstruct the image of the target at range R_0, the cross-correlation integral has to be applied, using the function for a single point target at R_0 for one of the functions in the integral and the synthetic aperture signal for the other. This procedure can be performed either by a digital computer[257] or by optical correlator. In the latter case holographic film is used to record the one-dimensional Fresnel-zone pattern.

Fig. 70 showed the basic layout of such a system if not continuous but pulsed sound is used for insonification. At each position of the transducer a signal is transmitted and the phase and amplitude information of the signal received in response to that transmission are received and placed into storage. Since the radiating element is moving in direction x, the storage of these signals resembles strongly those signals that would have been received by a one-dimensional linear array. Consequently, its operation can be regarded as that of a physical one-dimensional linear array with a large effective transducer-antenna aperture and this is the reason why sometimes this technique is called 'synthetic aperture' technique.

In 4.3.4 it was discussed that the linear array can be focused to a specific range. The same result can be achieved also by the synthetic aperture sonar by the proper adjustment of the phase of the received signal before summation. If the horizontal aperture of the transducer is D, its beamwidth will be given by the ratio λ/D, and it can be shown[257] that the linear resolution in azimuth δ_a is

$$\delta_a = \frac{\lambda R}{2} \cdot \frac{D}{R\lambda} = \frac{D}{2} \qquad\qquad 84$$

Equation 84 shows that the azimuth linear resolution is independent of both range and wavelength and, in contrary expectation, finer resolution is achievable with a smaller, not larger, physical aperture.

If for the storage of the signals photographic technique is used, then the simplest way to do this is (Fig. 71) to sweep vertically the electron beam of a cathode ray tube in synchrony with the transmitted pulse repetition rate. The vertical axis of the cathode ray tube will be the analog of the target range. This is imaged on photographic film which is driven in synchrony with the motion of the transducer. The obtained transparency will contain a series of one-dimensional acoustic holograms with the holograms running in the azimuth direction. When illuminating these transparencies with coherent light wavefront, we encounter the problem of the wavelength ratio of the insonifying and reconstructing wave. The azimuth and range coordinates of the insonified target are not in a single plane, and to superimpose them an aspheric lens system has to be used as shown in Fig. 72.

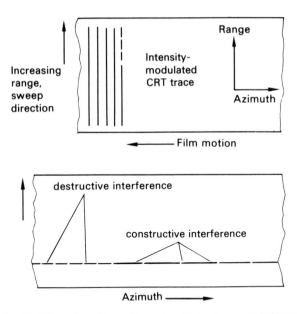

Fig. 71 Schematics of the writing out a side-looking sonar hologram.

Since it does not matter which part of a synthetic aperture system is moving with respect to the other, several types of side-looking sonar systems can be constructed for acoustic imaging depending on the task to be solved.

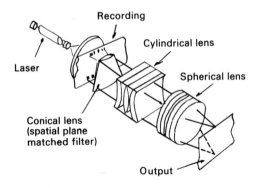

Fig. 72 Optical processor configuration for side-looking sonar signal processing.

6.1.3 Using sonic cameras

Two different types of sonic cameras for underwater viewing exist:

a) Using acoustic lenses to form the acoustic image.[259, 260]

b) Using no acoustic lenses but generally applying holographic techniques[261, 262].

160

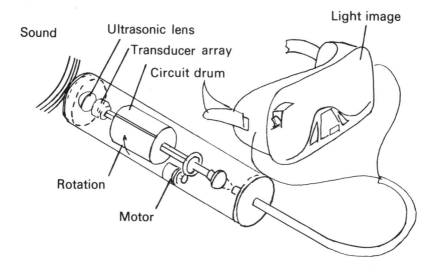

Fig. 73 Schematics of a head-mounted ultrasonic image converter.

6.1.3.1 Lens imaging A typical representative of underwater sonic viewing systems of group (a) is the head-mounted ultrasonic image converter developed by Rolle and Werle[260] (Fig. 73). A liquid acoustic lens focuses the sound image on a rotating transducer array consisting of 32 elements widely dispersed over the image plane of the lens, thus generating a 1000 element picture in real time. These elements are used both to transmit and to receive the sound beam. The signals from the array are

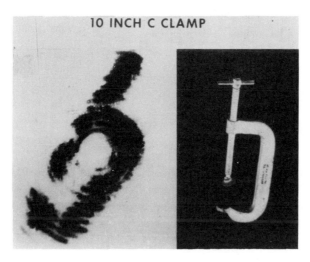

Plate 34 A 10 inch C-clamp seen through murky water from about 1 m distance with head-mounted ultrasonic image converter, and its optical image[260].

161

processed in a rotating electronic package and excite an array of 32 LED which are dispersed in the same format as the elements of the transducer array. The visual image from the LED array is conducted to a face mask by fiber optics and the diver views the image through combining prisms, which also permits the diver to see in the water.

This rotating scanning has also the advantage of dealing with specular reflection problems. Since only 3% of the total field of view is insonified at any instant, there are 97% fewer reflectors that can interfere or cross-talk into a particular beam which is coincident with a weakly reflecting portion of an object surface.

Plate 34 shows a 10 inch C clamp as seen through murky water from about 1 m distance.

Basically the same idea, only on a larger scale, is realized in the circular-scan camera developed at Westinghouse Research Laboratories. The schematic of its operation is illustrated in Fig. 74. A radial row of 128 transducers is rotated through 360° to obtain a complete scan of the circular area to be imaged with the aid of a sonic lens. The output of the amplifiers are connected to 128 light sources mounted on the top of a rotating cylinder. The output light pattern is imaged onto a photographic film by means of an acoustic lens to provide a visible record of the sound image. Plate 35 shows a cylinder with the conical nose when it was illuminated with a light source from below, while Plate 36 is its sound image at 3 MHz. The contour

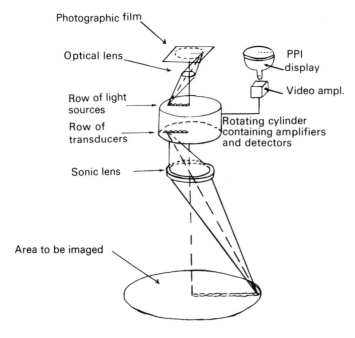

Fig. 74   Schematics of Westinghouse Research Laboratories' circular scan camera.

162

Plate 35 A cylinder with conical nose illuminated with light from below (*courtesy* of Westinghouse Research Laboratories).

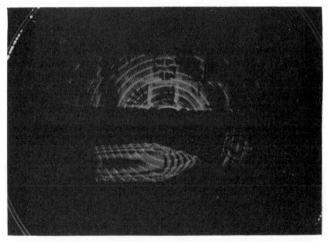

Plate 36 Sound image of a cylinder with conical nose insonified with 3 MHz and viewed with a circular scan camera developed by Westinghouse Research Laboratories (*courtesy* of Westinghouse Research Laboratories).

lines appearing on the sound image are the result of range-gating. The range contour changes from white to black with 1 inch range change each.

6.1.3.2 Lensless imaging Lens imaging is always limited by the aberrations in the lens, and this is especially true in acoustic imaging. Since holography does not require a lens to form an image, it was thought already at the early stage of acoustic holography that it could be used for underwater viewing. Since it gives range information too, it could eliminate all problems associated with moving acoustic lenses for focusing. The basic idea was to use a two-dimensional array to detect the acoustic waves reflected from the insonified target. By adding an electronic reference signal,

163

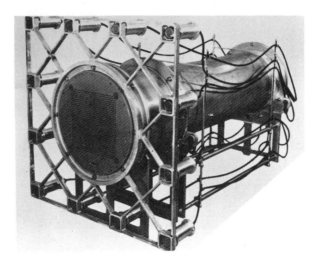

Plate 37   A 32 x 32 array sensor with 4 x 4 array of transmitters developed by OKI Electronics of Tokyo[267].

the resulting electrical signal, the holographic information could be processed in a conventional form to obtain visual image of the insonified scene. The soundness of the idea was first tested by Bendix Research Laboratories, and since the results have been encouraging, a worldwide activity was started to put the idea into practice. Plate 37 shows the system of OKI Electronics of Tokyo[267] which uses a 32 x 32 array of sensors with 4 x 4 array of transmitters, which means that it effectively synthesizes 128 x 128 sampling apertures. The resulting hologram is stored digitally

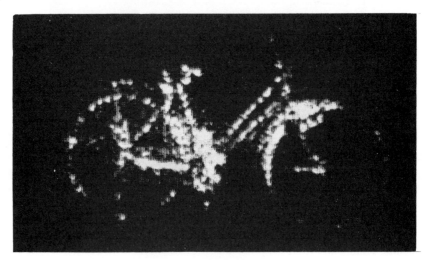

Plate 38   Visualized sound image of a bicycle under water[267].

164

and is processed by a minicomputer to yield the image (256 x 256). Plate 38 is the visualized sound image from a bicycle under water.

In underwater acoustic holography, beside the already mentioned problem, the biggest difficulty is perhaps that of achieving a versatile system for real-time reconstruction. Such a system has to fulfil the following requirements:

a) it must be sufficiently fast to meet the real-time processing requirement of 33 frames per second,

b) it must be reusable.

Although theoretically these requirements can be met by analog as well as by digital methods, practically the problem is not easy but it can be solved. The Westinghouse FFT3 cascade processor is capable of processing the hologram data of a 64 x 64 array in 6.14 msec providing ample time for real-time display of the reconstructed two-dimensional sound image. If, however, the hologram is first displayed on an array of light-emitting elements or CRT screen, no real-time reconstruction is possible since these are incoherent light patterns. A device is needed therefore which transforms an intensity pattern of incoherent light incident on the screen into a spatial amplitude and phase modulation of a visible wavefront, accomplished by the reflection or transmission of the reconstructing laser beam from or through them.

Farhat[264] proposed to use a gamma ruticon to reconstruct the image directly from the CRT screen. Boutin *et al*[265] developed an incoherent converting system based on the property of a thin-cut deuterated potassium dihydrogen phosphate (DKDP) crystal that its reflection index can be modulated by an electron beam; thus a coherent light transmitted through it will be phase modulated with the hologram information. A 2 x 2 cm area on this crystal can resolve about 300 lines corresponding to 9×10^4 data points which exceeds the present requirements of an underwater acoustic imaging system. However, there are still several technical problems to be solved. The crystal surface damaging effect of the electron bombardment has to be reduced, the outgassing from the crystal assembly which can poison the electron gun cathode has to be prevented, etc.

6.1.3.3 C-mode imaging Knowing that in acoustic holography the reconstructed image shows first order distortions especially in the frequency range used in underwater imaging, that is only two-dimensional images can be brought into focus, the rightful question arises whether it would not be more practical to use C-mode imaging. Problems related to attenuation and scattering of the wave during transmission through water or those associated with an acoustic lens could then be removed or substantially reduced. Multiplicative arrays of receiving elements are, however, unsuitable to apply to visualize the target, because they have no isotropic sonosensitivity to the sound field[266].

Shibate *et al*[267, 268] have recently shown that this problem can be solved by a coaxial-circular-spherical receiving array (CCS). Such an array consists of 24 receiving elements arranged in three concentric rings, each of which consists of eight elements[Fig. 75]. The eight output signals from the eight elements on a circlet are summed together as denoted by Σ on Fig. 75. The sound source is in the plane of the

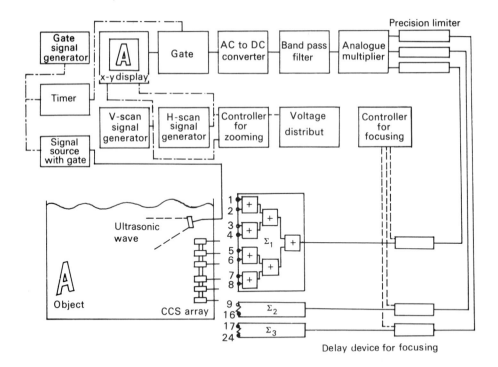

Fig. 75 Schematics of a coaxial circular receiving array[267].

array and emits the ultrasonic wave packet of about 500 microsecond duration at a repetition rate depending on the average target distance, e.g. 200 Hz. The array is focused by introducing a suitable delay into each of the additional output signals. The expansion of the image, that is zooming, can be made by changing the ratio of the voltage, led into the display unit for sweep to that lead into the voltage distributor for electronic scan.

6.2   Under earth

Geophysical exploration uses sound waves for a long time, but to extract an image-like information from the data as it is done in underwater viewing is far more complicated. In contrary to underwater viewing in earth the wave energy will be reflected and refracted not only by the object of interest but also by random variations in the medium. Consequently, the reflected sound energy will provide a mixed multicomponent signal comprised of object-representing and medium-representing components. As long as B-mode type (cross-section) images have to be formed, that is information on range and x direction is only required, this problem of origin is not so severe. They are also called reflection seismographs, and are an important tool of petroleum prospectors. They use many hundreds of surface source locations with

166

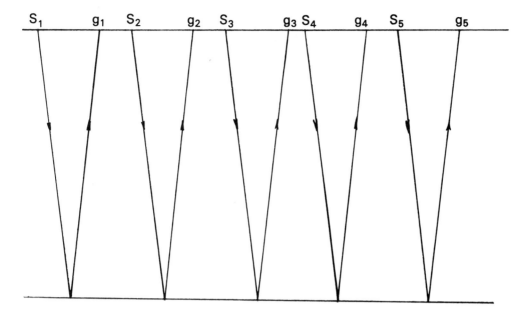

Fig. 76   Schematics of the sources and geophones location.

about 50 surface receiver groups, each consisting of 20 geophones locally summed for each shot point. Fig. 76 depicts this geometry with s and g referring to shot and geophone location. The frequencies of interest may be from 6 to 100 Hz. The reflection tomography is reconstructed by the computer. Fig. 77 shows such a typical sound image[269].

To use holographic technique for the determination of the orientation and shape of subsurface reflecting or refracting objects was first proposed by Silverman[270], and is called seismic holography. According to this, a steady state single frequency sinusoidal seismic wave on the earth surface has to be generated. The electrical signals resulting from the reflected waves recorded by a geophone spread should be combined with an alternating current wave of the same frequency. The resulting hologram pattern could then be reconstructed either by computer techniques or by optical means.

To avoid the problems associated with forming the communication path between the seismic source and *each* geophone, Farr[271] suggested to use the refraction path that extends from the sound source through the weathered layer to the first highspeed layer below, along it to the weathered zone below the geophones and up to the geophones.

It is not the aim of this book to explore in detail the technical feasibility of seismic holography, although the basic rules of it can be found in earlier chapters. Nevertheless, discussion of how to solve some special problems may be useful.

167

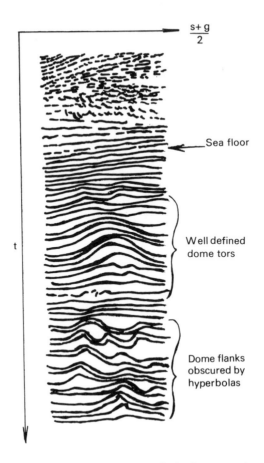

$$\frac{s+g}{2}$$

Sea floor

Well defined
dome tors

t

Dome flanks
obscured by
hyperbolas

Fig. 77   Computer reconstructed seismic tomography.

6.2.1   Bispectral holography

The disturbing effect of the heterogeneous medium can be considerably reduced according to Mueller[272], if two different frequencies are used for imaging. According to this suggestion[Fig. 78], a single lithologic variation in the form of a stratum of soil or a mask is disposed between the target and the earth surface. The sound source is situated on the earth surface approximately vertically above the target to be visualized.

The first recording is performed with a preselected frequency N_1, permitting penetration of the wave to the target, striking both, the object and the mask. This will reflect toward the earth's surface one object wave and a mask-reflected reference wave. Both are received by an array of seismic detectors. The reflected wave energy detected and converted to an electrical signal will be $S_1 + S_2$, S_1 being the signal

168

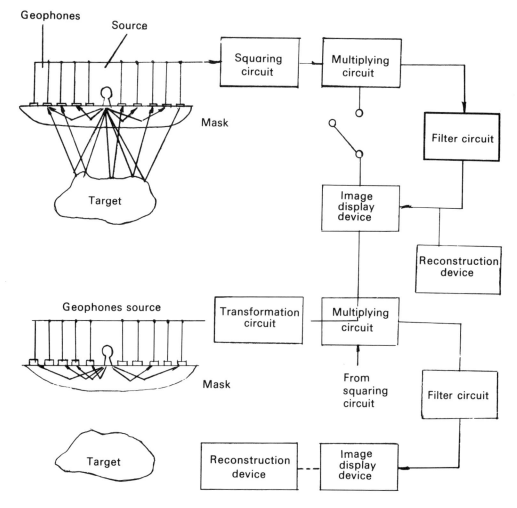

Fig. 78 Schematics of bispectral seismic holography, suggested by Mueller[272].

component of the target at frequency N_1, and S_2 that of the mask. In the second step only the mask is irradiated by a predetermined frequency N_2, taking into consideration that the higher the frequency the lower the penetration, and the detected electrical signal will be S_2, which is then transformed and processed with the formal signal $S_1 + S_2$ to provide an output representative of the object representing signal component S_1. The resulting signal then provides the sound image of the target.

Recently Sato and Sasaky[273] proposed another bispectral holographic technique to eliminate the image degradation effect of the random medium. Although this method is still in its very early stage, we believe that it may have a future in acoustic, especially in seismic imaging but less so in optics. The reason is simple. This method

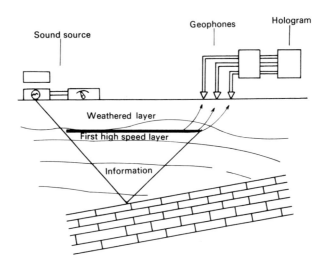

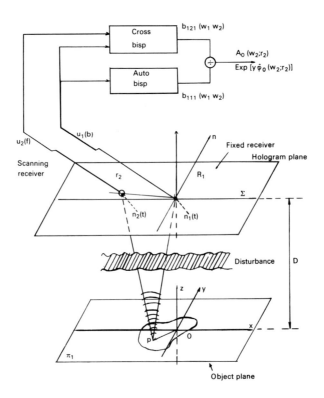

Fig. 79 Schematic construction and signal processing of proposed bispectral seismic holography.

170

requires detection of amplitude and phase, which can easily be met in acoustics and is rather difficult in optics.

The schematic construction of signal processing of this bispectral holography is shown in Fig. 79. It is assumed that the insonifying signal or those reflected from the target are random and non-Gaussian. This condition is generally satisfied, since any periodic signal can be regarded as a non-Gaussian signal.

The hologram is recorded by a fixed and a scanning transducer, by taking the ratio of the auto-bispectrum and cross-bispectrum of the signals detected. Although the strict theory states that bispectrum of a signal is expressed as a Fourier transform of the third order correlation function of the signal, in practice, the bispectrum can be calculated directly through fast Fourier transform of the signal, and so the calculation time is reduced to a practical level.

Image reconstruction can be performed by conventional holographic methods.

6.3   In spectral domain

This title may appear somewhat unusual at first glance, until we realize that physically natural information is always in a form of superimposed interferograms, and that our minds cannot handle such an interferogram. Only if it has been first transformed into the artificial information pattern known as *spectrum* can we relate it to the properties of matter, thus it will be clear what the title intends to express.

The information pattern called spectrum may convey two types of information. In the first type if informs about the wavelength dependence of a given property of the material, in the second type about the spatial structure of the material under investigation. Information belonging to the first group is bound to the amplitude of the information bearing wave, while those belonging to the second group are bound to the phase, or both amplitude and phase. Although a displayed spectrum is an 'artificial image', sometimes it gives more insight into a material than an image we consider to be 'natural'. There are several approaches how to see by sound in the spectral domain; the most promising attempts will be discussed.

6.3.1   Time-delayed spectroscopy

The technique of time delayed spectroscopy (TDS) was first described by Hayser[274], and later adapted for ultrasonic soft tissue visualization[275, 276]. It is a coherent information process in which both the time domain and frequency domain are utilized. The transmitted signal has a predetermined frequency spectrum with the equivalent of a time tag to each frequency component. In the simplest case this consists of a linear frequency sweep with time in which the tag is the moment of occurrence of each frequency. Upon emergence from the specimen under test the frequency components with a given time delay relative to the moment of transmission are reassembled to yield the delayed spectrum the process is named after.

171

This method not only provides the complete complex spectrum with amplitude and phase, but also signal components due to longer pathlength, such as those caused by scattering, are effectively suppressed because their spectrum time tags are rejected. In the case where time domain data are of significance, a real-time Fourier transformation can be made of this frequency spectrum. Recently, this method has been used to study blood coagulation mechanism[277].

A C-mode sound image can be generated with TDS if the ultrasonic transducers are synchronously scanned in a raster fashion while simultaneously deflecting the electron beam of a CRT. Only those sound waves which pass through the object with predetermined transit times are recorded; earlier or later arrivals which would cause image degradation are eliminated. TDS technique has the advantage that when the received signals are demodulated, the data give the amplitude and phase of the selected signal over the present frequency range. These give information on tissue attenuation and dispersion as a function of frequency. This technique works in transmission and reflection as well.

6.3.2   Ultrasonic holographic Fourier spectroscopy

The main area of Fourier transform spectroscopy is a relatively broadband spectrum. This is especially so when the spectrum of weak sources has to be investigated or if the sensitivity of the detector system is not quite adequate; this situation often occurs in the ultrasonic frequency range.

Although conventional Fourier spectroscopy could be applied to acoustic sources too, the precise moving of one of the reflectors of the two-beam interferometer causes several technical problems which do not make this method attractive at all to ultrasonographers. Since, however, it was first shown by Stroke[278] that a two-beam interferometer with tilted, instead of parallel moving, mirrors can record the Fourier hologram of an extended incoherent source of arbitrary spectral intensity distribution, there is interest to extend Fourier spectroscopy to acoustic radiation[279].

The basic layout of an acoustic holographic Fourier-spectrometer AHFS does not differ in principle from its optical counterpart (Fig. 80). The incoming ultrasonic beam is divided by a beam splitter. The two fixed mirrors are tilted in such a way that the image reflected by one forms an angle with the other so that the interference pattern of equal acoustic densities are localized at the bisector of this angle. Because these fringes are acoustic, their optical replica must be produced, if we wish to use conventional Fourier spectroscopy procedures combined with optical computing further on. There is no restriction as to what sort of visualization technique should be used, but it is preferable to apply one which uses coherent illumination to form the optical replica of the ultrasonic field. Thus the possibility of real-time optical Fourier transformation of the ultrasonic interferogram can be utilized. Plate 39 shows such ultrasonic holographic spectrogram as shown on an AOCC display. It demonstrates well that a spectroscopic hologram is an overlapping of sinusoidal gratings with spatial

172

frequency $\sigma\Theta_A$, where σ is the wave number and Θ_A is the angle between the tilted reflectors as indicated in Fig. 80. If the wave number of the reconstructed wavefront is σ_R, a part of the reconstructing wavefront will be deflected according to

$$\sin\alpha = \sigma\Theta_A/\sigma_R \qquad\qquad 85$$

With a lens of focal length F the distance of the first order from the optical axis will be

$$u = F\sin\Theta = \sigma F \Theta_A/\sigma_R \qquad\qquad 86$$

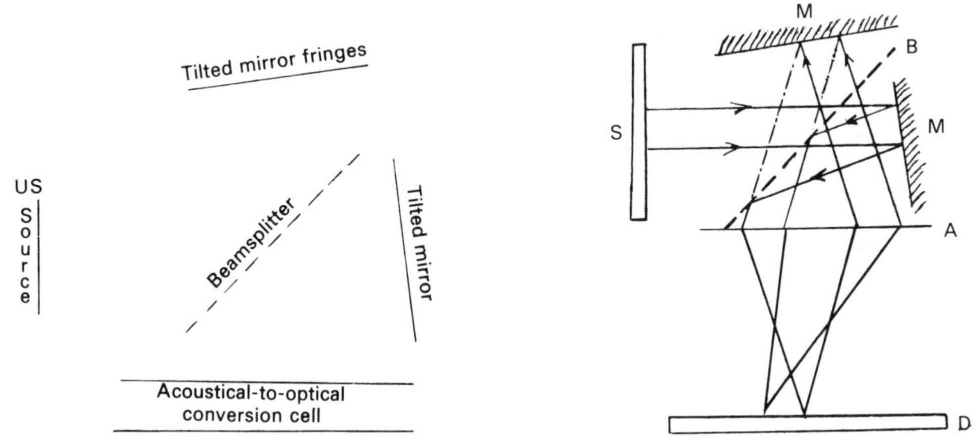

Fig. 80   Basic layout of an acoustic holographic Fourier spectrometer.

For optical computing visible light is used, thus σ_R will be several orders of magnitude larger than σ for both ultrasonic frequencies, and so we have the wavelength discrepancy problem already discussed. Fortunately, in this case, this is not as severe as in conventional ultrasonic holography, where three-dimensional visualization of the insonified object is of concern. It is somewhat similar to the problems faced in infrared spectroscopy and can be overcome by similar methods.

Ultrasonic holographic Fourier spectroscopy is still in its developing stage, and if really broadband transducers were available, it could help to solve problems of ultrasonic soft tissue differentiations; for example, discriminate between non-neoplastic, sterile inflammatory lesions and malignant tumors.

6.3.3   Back-scattering spectroscopy

As already discussed in 5.2 and 5.3, a B-mode and C-mode sound image is formed as the result of a sampling procedure in which the aim is to record the two-dimensional

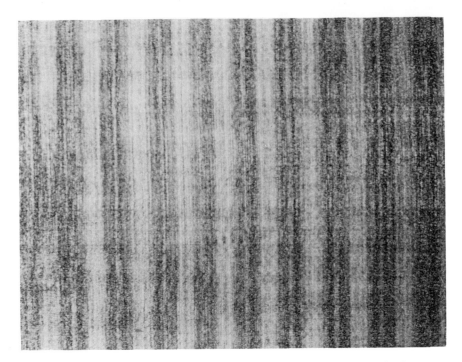

Plate 39   Ultrasonic holographic Fourier spectrogram recorded on an AOCC display[279].

matrix, the reflected sound packages, that arises from a set of geometrically well-defined volume element within the insonified target space. The value of the echo amplitude recorded from a given element will depend

   a) on the surface structure of the volume interface in question, R_s,

   b) on the reflective coefficient of the boundary due to acoustic impedance changes, R_{imp}, and

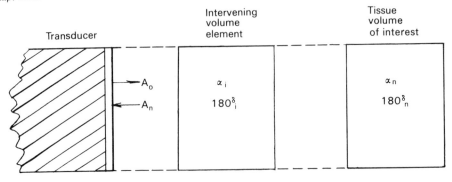

Fig. 81   Attenuation properties of the intervening elements α_i.

174

c) on the attenuating properties of the intervening elements, α_i, as shown in Fig. 81.

a) According to a semiempirical model of Senapati *et al*[280], for a low-loss medium the reflectance of a random rough surface may be defined by the ratio of the scattered to incident intensity, and can be given by

$$R_S = \frac{I}{I_o} = \frac{1}{\cos\theta\left(1 + \dfrac{k^2}{v^2 f^{2n}} \, \mathrm{tg}^2\theta\right)} \qquad\qquad 87$$

where Θ is the angle of incidence and v the rms of roughness. Both constant k and the frequency exponent n have to be determined experimentally, but in general, with a value n = 1.25 and k = 10^4 the theoretically and experimentally evaluated curves show rather good correlations, despite the simplicity of the model. They also demon-

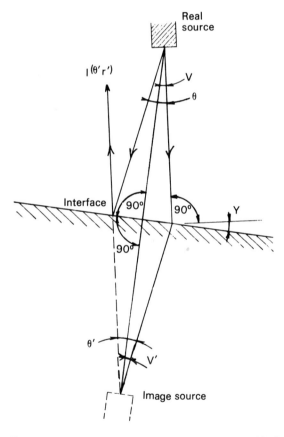

Fig. 82   The geometrical relationship between the real source and its image.

175

strate that there is a distinct increase in scattering intensity with frequency and similar angular dependence.

b) As already discussed in 3.5, the reflective coefficient may or may not be a complex quantity depending upon the acoustic losses in the two media. However, the power in the reflected wave is related to the power in the incident wave by the square of the magnitude of the reflecting coefficient R_{imp}, which is always a real quantity, i.e.,

$$W_R = R^2_{imp} W_1 \qquad\qquad 88$$

where W_R is the total reflected power, and W_1 the total incident power.

For reflecting interfaces which are not normal to the axis of the sound beam, the reflection coefficient is a function of incidence, but for small angles ($<5°$) this effect can be disregarded in practice. Assuming that the boundary conditions are such that image theory can be used, the original source and the reflecting interface can be replaced with an equivalent image source as shown in Fig. 82. According to Pinnel[281], the intensity of the reflected wave will be

$$I_R = \frac{W_0 \, |R_{imp}|^2 \, \exp(-2\alpha r_0) \, D^2 \, (\theta')}{r'^2} \qquad\qquad 89$$

where r_0 is the distance from the transmitter to the reflecting interface, r' and θ' are variables of a coordinate system oriented at the image source and related to the original variables (r and Θ) and to the orientation parameter γ.

As seen from Equation 89, to relate the signal detected by the receiver to the reflection coefficients and interface orientation is a rather difficult task, since it means finding an expression for the signal that is detected by the receivers with the image source as a transmitter. This requires the integration of the square of the directivity functions over the 'apertures' of the image source and the receiver. This is one of the reasons why back-scattering spectroscopy is still in its research stage.

c) The attenuation properties of the intervening elements can, however, be taken into account if the value from which the scattering is to be measured is well defined in shape and space. Pulsed continuous waves do permit this, but very short pulses are more advantageous for back-scattering spectroscopy since they have not only a very good volume resolution, but they have a broadband frequency spectrum too.

The radiation pattern of a transducer is, however, dependent on frequency and so the spectrum of radiation emitted by a pulse transducer will vary with position in the radiation field, this effect cannot be disregarded. Chivers, Hill, Nicolas and others[282-284], have studied this problem in detail, and have found that the back-scattered spectrum $S(f)$ for a volume element $1/2 \, c\tau$ where c is the sound velocity and τ the pulse width, is given by

$$S(f) = \epsilon \frac{S_m(f)}{R_0(f)} \exp(2x\alpha') \qquad\qquad 90$$

176

where S_m is the measured back-scattered spectrum; R_o the loop frequency response of the transducer; x the distance from the transducer to the element in question; α' the attenuation per unit length in the material; and ϵ is a correction factor including the effect of the surface structure and transmission across the interface.

As seen, the complexity involved in back-scattering spectroscopy is great and ambiguous. The relationship between the back-scattered signals and the structure of the back-scattering volume has not yet been established, but relations on a semi-empirical basis already allow the exploitation of this idea, especially if the spectrum may be deconvoluted by computer to obtain the relevant back-scattering cross-section.

This can be compared with less ambiguity to provide an empirical approach of describing the scattering material, e.g. for material testing or for tissue differentiation.

6.3.4 Recording ultrasonic spectrum analyzer

In many cases of scattering pattern analysis the Fourier transform description or frequency plane description is easier to interpret and manipulate to extract pattern recognition features than the corresponding spatial description since well defined relationships may be found between certain specific features and corresponding diffraction patterns.

A feature in pattern recognition terminology is a measurement which is later processed by a decision maker to provide output answers in a recognition problem. The key to the simplicity of this scattering pattern analysis is that if the information carrier wave is coherent, a simple focusing lens can be used to perform in real time such a rather elaborated computation as Fourier transformation. The one-to-one relationship between a function and its Fourier transform causes no loss of information if one uses the transform description instead of the spatial description of an input.

The acoustic space Fourier-analysis is most simply explained with the aid of thin lens and ray theory[285]. Let the focal length of the lens be F, and the coordinates chosen be such that the polar axis coincide with axis x and the plane of the lens with plane y, z. If a plane wave

$$\cos \left[\omega t - (\omega/c) \cos\Theta.x - (\omega/c) \sin\Theta.z \right]$$

inclined at angle Θ to axis x is impinging on an acoustic lens it will be brought to a point in the focal plane at distance $F \, \mathrm{tg}\Theta$ from the polar axis (Fig. 83). Similarly, if another infinite plane wave denotes by

$$\cos \left[\omega t - (\omega/c) \cos\Theta.x + (\omega/c) \sin\Theta.z \right]$$

impinges on the acoustic lens it will be brought to a point in the focal plane at distance $-f \, \text{tg}\Theta$ from the polar axis (Fig. 83b). Combining these two planes we get an infinite wave travelling parallel to axis x and processing an amplitude of

$$2 \cos \left[(\omega/c) \sin\Theta . z\right]$$

Thus, in the region of overlap, where the parts of the two plane waves of the same amplitude and frequency cross, the wave motion is indistinguishable from that produced by a single plane wave with 100% intensity modulation across the wavefront. This modulated wavefront can, however, also be regarded as the result of a wavefront that passes through a grating of spacing $1/2 \, \lambda \, \text{cossec}\theta$. The lens will resolve this wave into two focal points as shown in Plate 40. The distance from the polar axis can be expressed with good accuracy by

$$r = F\lambda k \qquad\qquad\qquad\qquad\qquad\qquad\qquad\qquad \textbf{91}$$

where k is the spatial frequency of the scattering structure, λ the wavelength, and F is the focal length of the lens.

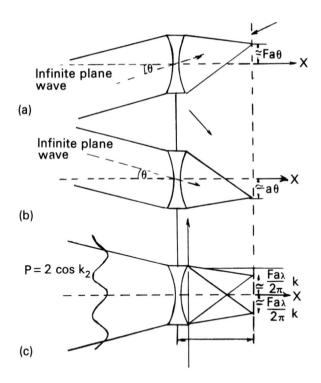

Fig. 83 Acoustic Fourier transform.

Plate 40 The acoustic Fraunhofer diffraction pattern of a grating.

Different names are used to describe the energy distribution in this plane, since intensities are measured, rather than amplitudes. This pattern is called modulus squared of the Fourier transform, acoustic power spectrum, Wiener spectrum, or Fraunhofer diffraction pattern. The power spectrum is centered on the polar axis; rotating the scattering media leads to a rotation of the spectrum which, however, remains centered on the polar axis. These characteristics of the acoustic power spectrum make it possible to detect or measure significant features of a material with

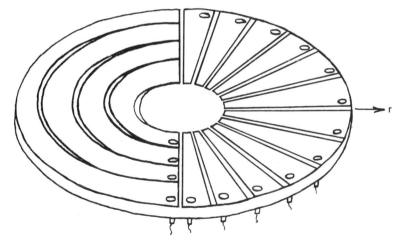

Fig. 84 Schematics of the detector array Recording Ultrasonic Spectrum Analyzer, RUSA.

179

a specified detector array, as first demonstrated in the acoustic domain by Greguss[286].

Fig. 84 shows a detector array which consists of parts of wedge-shaped elements and semi annular ring elements, which may be made from piezoelectric material, or be of condenser type, or foil-electret type.

This configuration allows the measurement of the energy distribution of the acoustic power spectrum. The use of polar coordinates leads to information about energy distribution in terms of spatial frequencies and azimuth; from zero in the centre the spatial frequency increases with increasing radius. The semi annular ring elements measure the average energy collected in each of the spatial frequency bands. This seems to be rather useful for the distinction of coarse and fine details of the insonified target. Finer details contribute more energy in the higher spatial frequency range. The wedge elements are useful for detecting energy spikes associated with sharp contour lines.

The acoustic power spectrum of the insonified target can be measured in transmission as well as in reflection if appropriate gating is used. Fig. 85 shows the schematics of a recording ultraconic spectrum analyzer RUSA which can operate in both modes. The detector array is placed in the focal plane of an acoustic Fourier lens, and since the zero order of the diffraction pattern is generally of limited interest, the detector array may have a hole in the center. This hole, however, acts as a spatial filter for the interrogating coherent ultrasonic beam generated by a curved transducer, the focus of which coincides with the hole and with the focus of the acoustic lens. Thus, the insonifying beam will be parallel when it impinges on the target.

The power spectrum of the back-scattered wavefront will be detected by a detector array and can be displayed on a CRT screen, as shown in Plate 41, or can be fed after digitalization into a computer for feature extraction.

The advantages gained from sampling the acoustic diffraction patterns are, first that the entire cross-section of the insonified sample contributes to the recognition

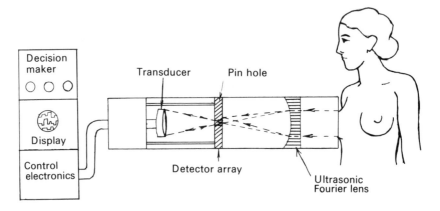

Fig. 85 Schematics of RUSA.

180

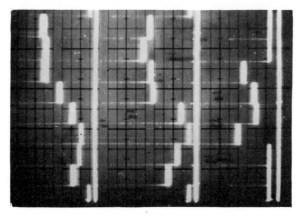

Plate 41 Acoustic power spectrum of a scattered wavefront recorded by RUSA and displayed on a CRT screen.

by modulating the energy content of each point in the diffraction pattern. Of course, this powerful parallel processing causes the useful acoustic information to be distributed over a wide dynamic range. Secondly, fewer frequency domain samples are needed to extract the significant features for the detection of inhomogeneities, as compared to spatial domain sampling. Furthermore, the detection of small irregularities is simplified, since they generate high spatial frequencies and so the corresponding frequency domain energy is rather diffracted from the origin and can be sampled by a large segment.

The RUSA serves as a processor which reduces an infinitely dimensioned pattern into an ordered set of n measurements, n being the number of elements. In the experimental device shown in Plate 42, n = 25. The role of the computer processing, then, is to implement a procedure which can assign the appropriate class label to an unknown sample presented by an n-dimensional vector. With proper feature selection

Plate 42 An experimental version of RUSA.

181

dimensionality can further be reduced, but sufficient information is retained to distinguish between classes.

The recognition of the target to be investigated is performed by a decision-maker which compares the feature read by the detector array with the features regarded to be characteristic of a known target. The success of the recognition depends, therefore, upon the features used which, as indicated, are combination of measurements. Naturally, several combination of measurements can be considered, and just to have an idea of the philosophy of selecting these features, let us assume that ring intensity patterns are recorded for a large number of specimens, and for each ring the maximum and minimum values are plotted. Naturally, no single ring will give a proper feature for recognition. However, we can try a feature vector defined as

$$F_v = \begin{bmatrix} \dfrac{x}{y} \end{bmatrix} \qquad\qquad 92$$

where y may be the sum of rings (n_1 - n_k), and x the sum of rings (n_m - n_z). If we now calculate $tg\Theta = y/x$, we get a single number as a one-dimensional feature. Most probably, a single feature vector will not be enough for correct decision making, thus a feature has to be derived by calculating the magnitude of successive differences between wedge outputs and then by saving the maximum of such differences as a feature.

Although the difficulties of obtaining good feature vectors are great, there is evidence from a limited number of measurements that there is a relationship between the structure of the material and the acoustic power spectrum. This technique will have practical value only if questions such as the effect of gate duration, the effect of the range in the material at which the gate opens, etc. are answered.

6.4   In nondestructive testing

As strange as it may sound, in the field where seeing by sound has had a continuously growing importance for nearly 50 years[5], in the field of nondestructive testing (NDT) of different materials[287] there is only a little, if any interest in sound images comparable to optical ones. Here and there suggestions are made to use B-mode[288] or C-mode[289], and quite recently Bragg[290] and holographic[291] imaging instead of A-mode display, but their general use, at least at present, is practically out of question. The reasons for this are manifold.

First, the sound frequencies used in NDT allow only the imaging of rather low spatial frequencies, and even if utilizable sound images are attainable by one or the other sound imaging techniques mentioned, the cost of instrumentation is, in general, not in proportion to the plus information obtained. Nevertheless, there are some new attempts which are worth mentioning.

Secondly, in most cases the shape of the flaw is less important than the knowledge of its location, and which type of flaw it is. A-mode techniques give the answer to the question of location, and even to the question of material structure by using ultrasonic spectroscopy, as demonstrated by Papadakis[292].

Ultrasonic spectroscopy is especially useful in seeing the texture in worked metals and other polycrystals. It is based on the fact that working operation on metals gives the polycrystalline material a crystalline type macrosymmetry resulting in an ultrasonic double refraction. A transverse elastic wave propagating along one of the three

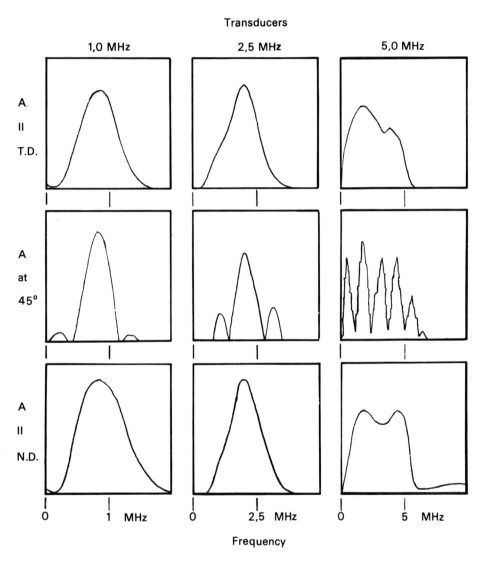

Fig. 86   Echo pattern due to ultrasonic double reflection.

orthogonal characteristic directions in a specimen, and polarized at an angle between the other two characteristic directions, is resolved in two waves polarized along the latter two directions and propagating at velocities differing by Δc. If the angle is 45°, the two waves of a given frequency will be by $(2n - 1)\pi$ out of phase, where n is a positive integer, and when recombined, will result in a vector rotated 90° from the original polarization, and thus cannot be detected by a receiver polarized parallel to the original transmitter. If ultrasonic pulse echo technique is used, the result is a series of nodes of amplitude within the echo pattern, as shown in Fig. 86. From the frequencies of the nodes the fractional velocity difference $\Delta c/c$ can be computed and related to the texture.

6.4.1 New approach to C-mode imaging in NDT

C-mode sound imaging in NDT is made difficult by two basic requirements. First, liquid coupling is required and the scanning transducer must follow the specimen surface contours with the sound beam axis normal to the local tangent of the surface, and secondly, the duration of complete specimen scan could be considerable. Recently, a new concept was introduced by Curtis[292] which allows real-time through-transmission C-mode imaging, and does not have these two basic requirements.

This method is based on the idea that a 25 micrometer thick polyethylene tetraphthalate film metalized on one side can be used on a conducting substrate to produce an ultrasonic capacitive transducer of practical sensitivity and broadband-width. The light flexible film under the influence of a DC voltage applied to the metalizing adheres to a conducting surface by electrostatic attraction, trapping pockets of air beneath it. (Plate 43).

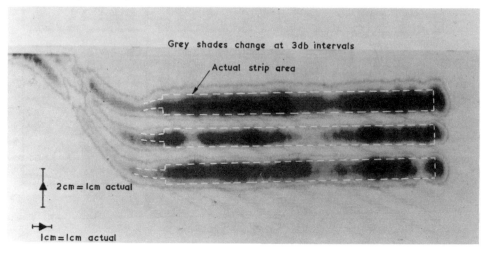

Plate 43 Beam plot from three polyscan receiver strips connected in parallel[292].

184

The air-film forms the active component in the capacitor formed with the metalizing as one electrode, and the substrate as the other. Forming transducers of any shape, size, or distribution together with their connection is easily achieved by photographic etch techniques. Although a pair of transducers working in a through-transmission mode have a joint sensitivity of 30 dB less than a pair of conventional damped PZT probes, this sensitivity reaches the sensitivity requirements for most nondestructive testing tasks.

Fig. 87 shows diagrammatically the basic principles of the new technique. Two rectangular transducers are arranged at right angles to one another. One is used to send a pulse of sound across the space between it and the second transducer which acts as a receiver. The acoustic beam profile of each transducer, assuming that they behave as pistons, is a single lobe. The cross-combination thus isolates a volume of space which is approximately a parallelepiped, if the wavelength of the ultrasound used is small as compared to the width of the transducer. Adding further (M-1) receiver strips adjacent to the first, yields a linear array of M through-transmission transducer path from one transmitter strip. If other (N-1) transmitter transducer strips are added, as shown, in Fig. 88, the crossed arrays form an N x M set of single transducer pairs. If each transducer strip is pulsed M times, and the M receiver strips are sequentially scanned, then, a through-transmission C-mode image is formed of the volume enclosed by the two arrays.

Plate 44 shows such an array of 32 transducer strips adhering electrostatically to the thin metal wall of a test tank. The transmitter arrays are placed on the opposite wall with the strip axis 90° to the receiver array. The box on the left contains the 32 amplifiers and MOS switches associated with the receiver array.

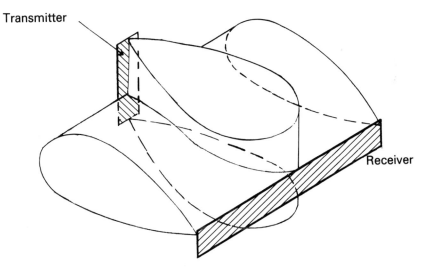

Transmitter

Receiver

Fig. 87   Basics of POLYSCAN through-transmission C-mode imaging.

185

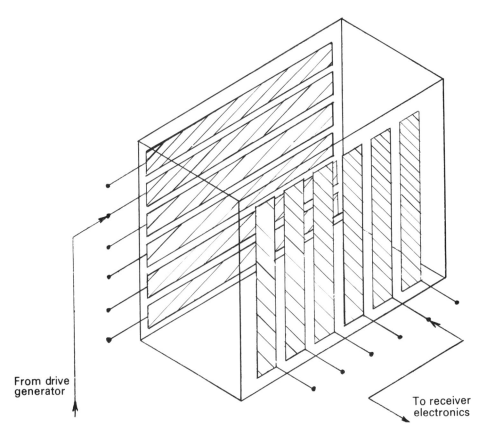

From drive
generator

To receiver
electronics

Fig. 88   Schematic arrangement of transmitter and receiver strip arrays.

A small computer, for example, PDP 8/E, is used to control both the transmitter-drive/receiver scan process and to form the C-mode image, which is produced upon a CRT screen as a checkerboard of N x M squares. The computer causes a grid of bright dotted lines to appear, representing the checkerboard matrix. At this point, the computer compares the transmitted amplitude value which has been stored for the first transmitter/first receiver combination with the number entered by the operator. This number represents a discrimination level that can be used to assess spatial variation in transmitted amplitude. If the stored value is greater than this, the computer fills the first square with bright dots. If the value is less, the square is left blank, and represents zero sound transmission through the test object. The computer then fills or leaves the next square blank according to the relative value of the sound transmitted between the first transmitter and the second receiver, and so on, as long as the C-mode image is not fully displayed on the CRT screen.

Although this technique called *Polyscan* suffers at present from several limitations,

186

Plate 44 Polyscan real-time through-transmission C-mode imaging system[292].

such as variable and for some purposes low sensitivity of the transducer array, and the difficulty in making electrical contact to the metalized foil, it shows great potential in becoming a versatile sound imaging technique for any constructive solid with a pair of parallel curved surfaces which can be contoured by the polymeric foil arrays. The technological difficulties mentioned can be surely overcome.

6.4.2 Passive sound imaging

All of the sound imaging techniques discussed so far have in common that the target had to be insonified by an external and independent source in order to get a sound image. They are therefore referred to as active imaging methods. The counterpart, passive sound imaging, i.e. the imaging of sound sources as a parallel to the imaging of self-luminous objects in optics, has not yet been intensively investigated, since there was no real demand for it. In nondestructive testing, however, there is a growing interest in the subject: acoustic emission analysis (SEA) has started to become popular.

Acoustic emissions are elastic waves produced by deformation taking place in stressed materials. They are generated by the conversion of stored elastic energy into kinetic form. The sound source may be the microscopic strain field surrounding a dislocation or the macroscopic strain field which drives a crack forward in a stressed

187

structure. The characteristics of the emission reflect in detail the nature of the deformation taking place. Because emission depends strongly on the material, only rules of thumb can be given about what type of information is conveyed by which parameter of the detected wave. So, information on

a) time-structure of source event is carried by the waveform,

b) energy of the source event is carried by the amplitude of the wave,

c) nature of the source mechanism is carried by the amplitude distribution,

d) nature of the source event is carried by the frequency spectrum,

e) shape and source location is carried by the relative arrival time at several observation points.

Recently, Curtis *et al*[294, 295], have shown that when applying polymeric foil transducer to the bound area of a metal–metal adhesively bonded lap joint or on a carbon fibre reinforced plastic CFRD fracture, started emission can be mapped of. The half-tone presentation of the emission spectra helps to differentiate between the fibre fracture and those for interlaminar fracture. Plate 45 is a typical example for the conventional and half-tone presentation of the frequency compounds of acoustic emission. The greater the degree of blackness the more energetic the spectra components are.

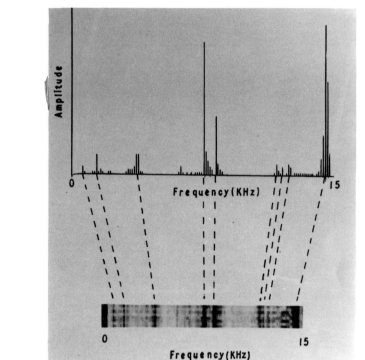

Plate 45   Half-tone presentation of the frequency components of acoustic emission[299].

188

In connection with SEA we have to call the attention to the already discussed bispectral holographic technique of Sato *et al*, which may provide another approach to record, for example an image of the sound emitting crack. The emitted signal could be used as the object wave, and the circumstances surrounding the crack may result in comparatively large Gaussian background noises, although a suitable combination of frequencies which satisfy $N_1 + N_2 + N_3 = 0$ should be selected beforehand from the signals.

6.5   In medicine and biology

The introduction of seeing by sound in medicine started in 1937 when the Dussik brothers[296], first transmitted a homogeneous, continuous ultrasonic wavefront

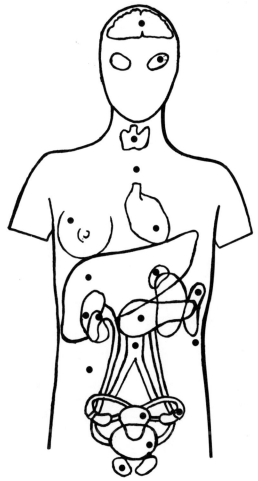

Fig. 89   Areas where ultrasonic imaging techniques can be used for medical diagnostics.

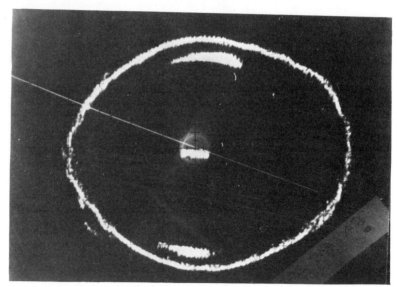

Plate 46   B-mode sound images of four horizontal cross-sections of the head obtained at different levels varied by 1 cm[301].

across a head, and recorded the intensity distribution on the other side of the head, which they called 'hyperphonogram'. The ultrasonic energy interacted, however, at every interface encountered, so that the emerging beam consisted of energy scattered by the summation of its interaction with every interface throughout its propagation through the head in an unknown and uncontrollable way. Thus, the two-dimensional information pattern recorded could not be unambiguously interpreted. For this reason

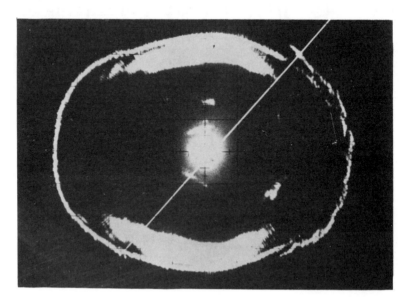

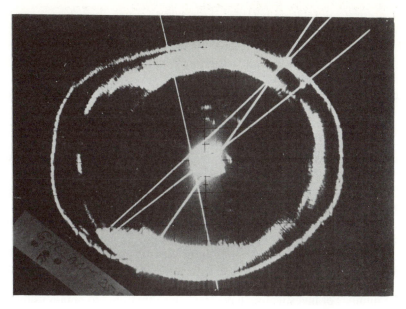

endeavours to use sound sources for imaging in medical diagnostics had been abandoned for more than a decade. In 1949 Ludwig and Struthers[297] used impulse echo technique to see gallstones and foreign bodies in tissues. Fig. 89 shows the area which can at present be seen by one or other ultrasonic imaging technique. There is tremendous literature on the subject of ultrasonic diagnostics[297–300] and therefore, we review medical sound imaging areas only so the reader may get a feeling of the

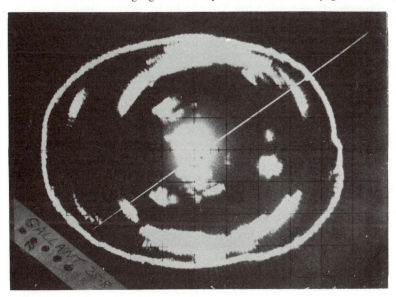

trends at present, and decide whether these trends are really to be followed or new directions should be encouraged.

6.5.0.1 Sound imaging of the brain. Though the Dussiks were not successful in imaging the brain, Makow *et al*[301] tried in 1964 to obtain a cross-section by using compound scanning technique, they combined reversible scanning and an outer movement around the head. Plate 46 shows four horizontal cross-sections obtained at different levels varied by 1 cm. These B-mode sound images have a far better quality than the hyperphonograms the Dussiks had, nevertheless, their information content is not enough, or not readable enough for anatomical and clinical purposes. The reason for this is, as discussed by White *et al*[302], that the skull causes topographical mislocation in the range of the image of the brain due to changes in sound velocity through its varying thickness, and there is also a mislocation in azimuth by reflection. Similarly, mislocation in azimuth may result from echoes of the central axis of the transducer but far away to sonotransparent areas of the skull, being imaged as if they lay in the central axis.

There are problems concerning resolution too. Because there is a varying attenuation at different parts of the skull, which causes uniform point reflectors to return different echoes sometimes, there will be a resolution degradation in azimuth.

A large number of investigators have tried since that time to obtain usable B-mode sound images of the head and they could well demonstrate the cerebral midline, however, no reliable representations of the intracranial structure have yet been achieved by sound imaging. Therefore, the present belief is that there is little hope to find a sound imaging technique which could do this job.

The author is, however, convinced that there are several sound imaging methods which should be investigated whether they could give the answer to this problem or not. So, ultrasonic transmission tomography by reconstruction discussed in 5.4.5.2 or

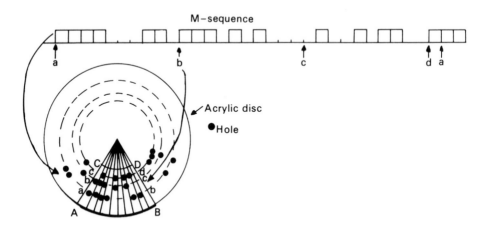

Fig. 90 Schematics of an ultrasonic imaging system using rotating M-sequence phase disc and correlation analysis after Sato *et al*[303, 304].

192

bispectral holography discussed in 6.2.1. could be prospective candidates for this task.

An ultrasonic imaging system using a rotating M-sequence phase disc and correlation analysis has recently been proposed by Sato *et al*[303, 304]. The random phase mask has holes according to the 1 of a M-sequence along circles. The same sequence is used for circles of different diameter in the disc with the corresponding different initial positions as shown in Fig. 90. Hence, in the region ABCD of the disc different portions of the M sequence are arranged and the thickness of the disc is chosen so that B-phase difference is produced between the waves that have passed the disc and those that have passed through the holes directly in the medium.

The disc is rotated at a constant rate and the cross correlation between the randomly phased mask and the original M sequence is taken. It is clear that only one point in the region ABCD has sharp correlation with the M sequence for a fixed delay of the cross correlation. Therefore, if this mask is placed behind the object, that is, if the object wave field is coded by this random phased mask, the cross-correlation function between the resulting field and the original M sequence gives the distribution of the wave-field. The schematic construction of such an imaging system can be seen in Fig. 91.

Only a fixed point receiver is used to detect the transformed field, and the cross-correlation function between this detected signal and the M sequence generated by logical circuits is calculated in a mini-computer. The result is distributed in the corresponding region ABCD on the image, according to the relation of Fig. 90, and the image is displayed on a CRT tube. Plate 47 shows the sound image of an 'L'-like object.

Until the previously discussed problems of sound imaging of the brain are solved, seeing by sound into the brain will mainly be restricted to the A-mode imaging of the middle-line. The conventional middle-line echoencephalography, however, achieves acceptable accuracy by means of the clinical bias of the operator and interpreter. This means that the measurements are biased and influenced by the operator, and this may lead to false negative errors[305]. By combining mid-line echoencephalography with computer technique this problem can be solved. In this

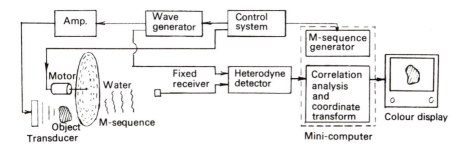

Fig. 91 Practical application of Sato's technique.

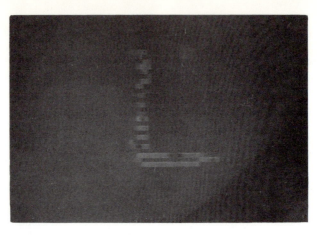

Plate 47 Sound image of an 'L'-shaped object recorded by a rotating M-sequence phase disc and correlation analysis[303].

case the transmit search pulse triggers a timing-clock. Echoes are amplified and gated into two memories called distant and mid-memories. These memories are interrogated by the pattern recognition circuit which either starts the computer or resets the whole system, depending upon whether or not the echo patterns received were acceptable.

Upon receipt of a correct pattern of distal and midechoes, the computer performs all of the arithmetic operations necessary for the determinations of the displacement of the mid-structure from the central line of the head. The distal and mid-structure echo groups are then analyzed on the basis of the number of individual echoes in a group on amplitude and sign of arrival.

The basis of acceptance of the distal echo pattern is the receipt of an echo doublet that extends a fixed amplitude discrimination level and arises in an interval gate that is open approximately 10 cm beyond the transducer. The round trip time for the edge detector to the skin-air interface is stored digitally as a number of timer counts. In place of the CRT screen, all information about the progress of the determination is fed back to the operator by audi signals; and simultaneously, it institutes a search for echoes in the mid-head region. Thus, every mid-line determination is normalized to the exact total head size at the location and orientation of the detector beam at that search time. If an acceptable middle-echo pattern is not received within the same 2 msec period as the distal, the whole memory is erased and the total determination begins again.

The brevity of this total research period ensures that the mid-line measurement will always be normalized to the distal echoes, since probe movements in 2 msec could not possibly cause significant relative misalignment of the mid- and distal-echoes[306].

6.5.0.2 Sound imaging of the eye   Ultrasound has proved to be of unique value in diagnostic tests of the eye region[307, 308]. A-mode and B-mode imaging are easily performed and especially B-mode imaging is useful in evaluation of adjacent

194

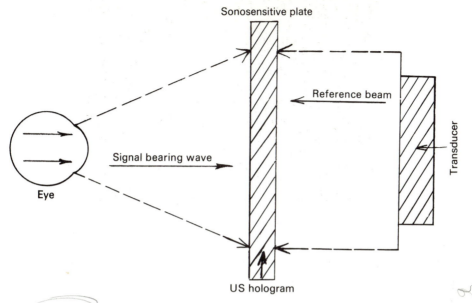

Sonosensitive plate

Reference beam

Transducer

Signal bearing wave

Eye

US hologram

Fig. 92 Schematics of the first experimental setup to record ocular sonoholograms.

orbital tissues[309]. Since the eye is one of the simplest anatomical structures from the point of sound imaging, it is not at all surprising that when acoustic holography was first thought of to be used in medical diagnostics, it was the eye where the idea was first tried[310]. The schematics of the experimental set-up is shown in Fig. 92. A piezoelectric transducer sends a parallel beam into a tube which contains the necessary liquid to record the hologram on a sonosensitized plate, which is placed between the transducer and the eye. That part of the ultrasonic beam that passes through the plate acts as a reference beam, while the portion of the beam which was

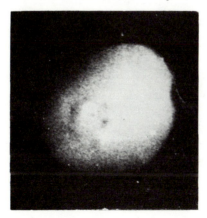

Plate 48   One of the first reconstructions of an ophthalmic ultrasonic hologram dated 1965.

195

reflected from the eye impinges on the sonosensitive plate forming an in-line holo-gram with the reference beam. After developing the plate, it can be reconstructed in coherent light, but the problems of wavelength discrepancy discussed in 5.4.1 have to be taken into consideration. Plate 48 shows such a reconstruction.

The relatively long exposure time required and the lack of easily interpretable reconstruction have discouraged the further development of this method, especially when the B-mode imaging techniques started to provide an almost complete ultra-sonic armoury for ophthalmic imaging and the interest was turned toward the interpretation of the resulting sound images.

But just this latter demand initiated the re-examination of ophthalmic holography. If, with improving technology, a hologram could be recorded for the whole globe and orbit, the ophthalmologist would have the hard copy of information from a three-dimensional volume, which then he could study at any time to extract information for differential diagnosis.

At present, the synthetic aperture method seems to be the best approach to solve this problem, since it is basically a B-scan system, as discussed in 6.1.2. Chivers[311],

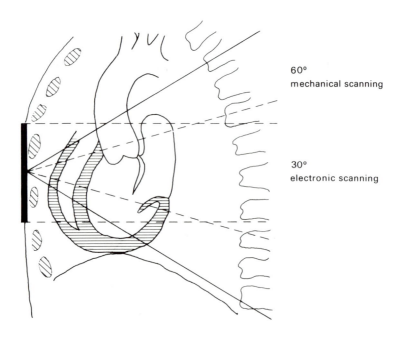

60°
mechanical scanning

30°
electronic scanning

Fig. 93   The viewing area of mechanical section scan and electronically scanned systems.

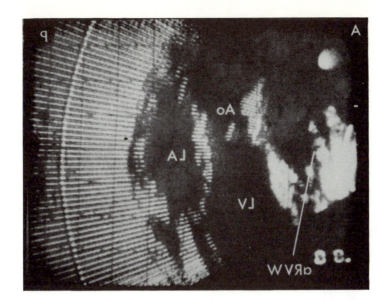

Plate 49   Longitudinal view through the aortic route of left ventricle from an infant with a patent ductus arteriosus[315].

Aldridge *et al*[312] succeeded to develop an opthalmic acoustic hologram scanner, which needs only 23 sec to scan the hologram aperture. When the problems of conflicting requirements concerning the transducers for B-mode and holographic

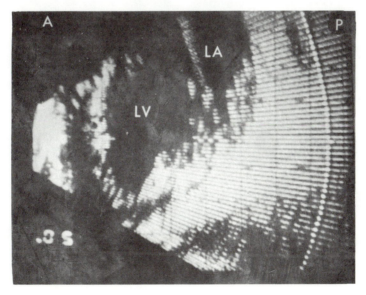

Plate 50   The same as Plate 49, only the scanner was tilted inferiorly.

imaging is solved—ophthalmology may become one of the rare medical fields where acoustic holography has a future but good holograms require strongly focused, short focal length transducers with long pulses, while B-scans need weakly focused transducers with long focal length and short pulses.

The investigation of the frequency domain of ophthalmic ultrasound is starting to become a versatile tool in the differential diagnostics of the lesion of the eye as it has recently been well documented by Trier[313].

6.5.0.3 *Sound imaging of the heart* The currently used one-dimensional echo-cardiograms (M-mode images) which depict range as a function of time, although very useful, have several limitations which can be overcome only if two-dimensional B-mode imaging of the heart in motion could be recorded. Several techniques have recently been suggested[314–318] based on the various mechanical and sampling techniques discussed in the previous chapters but it seems that no absolute technique exists yet. The competition at present is between the mechanical section scan and the electronically scanned system. Fig. *93* shows the viewing area of these scanners.

Since sound imaging of the heart of an infant is far more difficult than that of an adult we wish to demonstrate the state of art in intra cardiac sound imaging with Plates 49 and 50. The scanner has to have a precise lateral resolution, clear margins in the near-field and a scanning angle sufficiently large to permit accurate assessment of the relationships between different intra cardiac structures. Plate 49 is a longi-tudinal view through the aortic route of left ventricle from a child of 28 kg with a

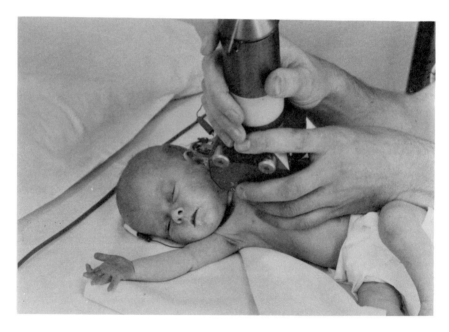

Plate 51 Scanner developed by Shaw *et al.*

198

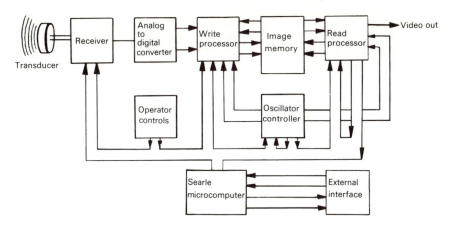

Fig. 94   Coupling existing sound imaging techniques with minicomputers.

patent ductus arteriaeosus, while Plate 50 shows the longitudinal view from the same chest position but with a scanner tilted inferiorly to show the apex of the left ventricle. These sound images were recorded with the sector scanner developed by Shaw *et al* which was placed on the chest of the child as shown in Plate 51.

6.5.0.4 Sound imaging of abdominal, intrauterine and fetal structures   At the present state of sound imaging the problem of getting a clinically usable sound image of abdominal intrauterine and fetal structures is in the first place not a technical one, but rather that of the interpretation of the obtained sound image. As was seen previously, the factors influencing the final visible appearance of the sound image are rather well known, only they are so numerous that it is difficult to account for them in the evaluation of the displayed sound image. Therefore it is not at all surprising that not only the researchers, but manufacturers also investigate the possibility to couple existing sound imaging techniques with micro-computers to account at least for some of these factors.

Fig. 94 illustrates the idea of one of these attempts. It shows an ultrasound transducer receiving two echoes from a given interface in a body, while the transducer is at two different positions. Applying the peak detection technique, the memory of the microcomputer stores the largest signal so that the visualized sound image intensity at this point will be I_1 if $I_1 > I_2$, or I_2 if $I_2 > I_1$. For instance, the Pho/Sonic-SM system uses this technique to reduce the artifacts caused by scanning speed and to obtain a more reliable gray scale B-mode sound image. Plate 52 is a photograph of such a recording[319].

In conventional B-mode imaging, which is generally used in the sound imaging of space occupying structures, the scanning transducer has two translational and one angular motion in a single selected scanning plane. Since, however, the operator and the patient are in a three-dimensional environment, the scanning aperture, the transducer should have the capability of being placed 'in a controlled way' in contact with

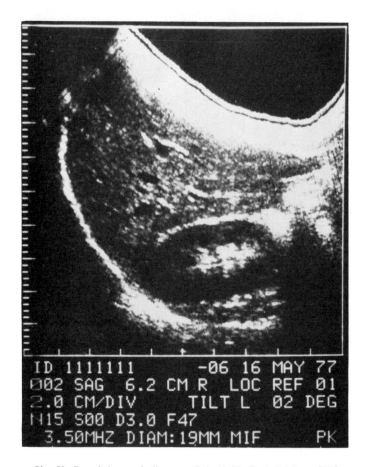

Plate 52 B-mode image of a liver recorded with Pho/Sonic-SM System[319].

any part of the patient's skin and to direct the insonifying beam in *any* direction into the patient's body from the point of contact, carrying out *any* scanning pattern the operator thinks appropriate. This idea has been put into practice by Brown *et al*[320] in the system called multiplanar scanning system and has been realized by Sonicaid Ltd. As shown in Plate 53, the transducer pivots freely around the two axes 1 within the scanning head. The transducer scan swings longitudinally or transversely and by combination in any intermediary direction. The scanning head is at the end of the boom which is counterbalanced and mounted on the measuring cabinet 2. It is free to move laterally or vertically 3, and to rotate about its own axis 4, turning the scanning head with it. The full three-dimensional freedom is provided by the measuring cabinet movement mounted on the support column moving lengthwise within the base of the scanner.

A three-axial measuring system monitors the transducer position in any direction,

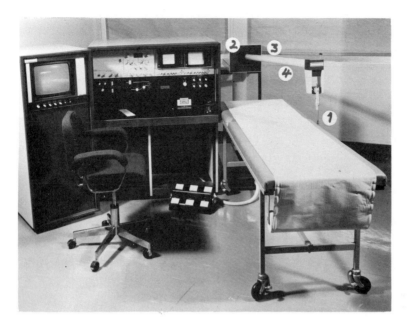

as a result of which the position of every echo received can be uniquely defined in three-dimensonal space, without the need for digital computing facilities.

A viewing control unit allows the operator to select the viewpoint from which the ultrasonic image is presented before the scanning commences. The resulting image can then be displayed from that viewpoint either in monocular form or as a pair of stereoscopic images giving the viewer the feeling of three-dimensionality.

The viewpoint of the observer can be rotated electronically to an oblique position, or a number of transverse sections at different levels can be superimposed on one picture, whereby they appear in their correct relative positions somewhat similar as in isometric representation of an object. Plate 54 is such a composite scan of a fetus.

The multiplanar scanning system identifies the x, y and z coordinates from each received echo, from each pixel, and so there is a theoretical possibility to use these data to print out a hologram from the insonified volume. When serving as a display, this would reconstruct the three-dimensional virtual image of the insonified body volume, somewhat similarly to that suggested by Greguss for holographic displays of computer assisted tomography[321]. It can be assumed that a plane exists at a distance z_1 from the recording transducer, where interference between the wavelets issuing from this pixel and a reference beam is displayed, and further that the 'nondisturbed portion of the beam' and that issuing from a given pixel are in phase at a given $x = O$, $y = O$ point on the display plane. Theoretical analysis reveals that after some simplifying assumptions[322] a binary hologram can be formed. The determination

201

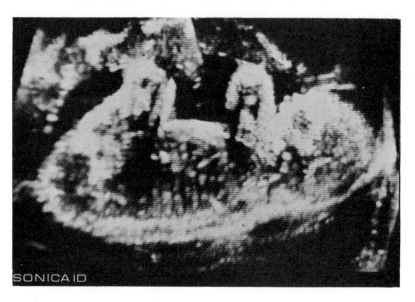

Plate 54   Composite scan of a fetus (*courtesy* of Sonicaid Limited).

of the portion of such a generalized holographic display system that should be opaque or transparent could be achieved simply by investigating the sign of the function where k is the wave propagation number.

$$F(x,y) = \sin \frac{k(x^2 + y^2)}{2z_1} \qquad\qquad 93$$

If this idea really works, it has to be proved, but there seems no reason why it should not if

a) the plane chosen is parallel with at least one of the scanning planes,

b) its distance from the body volume is chosen so that the aperture of the display allows parallax perception.

In realizing this concept some technical problems, such as the accurate plotting of the computer hologram and its reduction for light wave reconstruction have to be solved.

6.5.0.5 Sound imaging of space-occupying lesions   From the shape and size of an organ, information on its normal or pathological function can be drawn. However, in some cases of space-occupying lesions, a more detailed and a more easily interpretable information pattern for differential diagnosis is needed than a sound image which provides only simple outlines or contours. This is one main reason why gray scale sound imaging is becoming more and more popular, since it is thought that gray scales contain the extra needed information. This is true, however, only if there is an *a priori* knowledge of the interaction of the sound wave with the lesion under investigation and its surroundings, and if the visual sound image really displays this

202

information. To demonstrate what is really meant by this statement, we discuss briefly the problem of the differentiation of a benign breast lesion from a malignant one.

The efforts to differentiate breast lesions has a history of more than 25 years[323] and a good review on the success achieved so far can be found in the paper of Kobayashi[324].

One cue for differentiation can be obtained, naturally, by analyzing the shape and boundaries of the visualized patterns, since malignant lesions are usually irregular, and not smooth in outlines. But this is not enough information for differentiation, when conventional bistable B-mode imaging is used, since it is difficult to define what is irregular and what is smooth. However, it is known that malignant tumor tissues have higher sound absorption than benign lesions. Thus echo signals from the distant boundary limits of malignant lesions may disappear completely when the sound image is recorded at low sensitivity. At higher sensitivity setting the distant boundary will also show up in the case of a benign lesion.

To be able to differentiate between the two cases, the shape of the "shadow" behind the space-occupying lesion has to be displayed in such a manner that it can be determined whether it is homogeneous or has a swallow tail form. In the latter case, the lesion is nonmalignant, because the pattern is a result of a 'ringing' effect arising from a low energy loss as the beam passes through the distal lesion wall leaving sufficient energy to cause multiple reflections between the distal lesion wall and the chest wall, and the result of the total nonspecular reflection of ultrasonic energy at the lateral walls of the benign lesions.

As long as in bistable B-mode imaging this type of information can only be obtained by changing the sensitivity of the display and by recording several sound images. With appropriately programmed gray scale imaging all this information can be obtained on a single sound image. Further, gray scale sound images may display the echoes within the tumor with a higher fidelity than bistable imaging does, and this may lead to further differential cues. However, this is also the very point where great mistakes can be made, since our general knowledge on the ultrasonic properties of these tissues is very limited.

Lutz and Petzoldt[325] recently reviewed the possibilities and limitations of imaging space-occupying lesions, and concluded, among others, that

a) thyroid gland disorders, such as parenchymatous stroma and nodual goiter can be differentiated, but no distinction can be made between toxic adenoma and malignant tumor;

b) pancreatic disorders can be differentiated if the lesion is not too small;

c) gall bladder disorders as cholelithiasis and papillomas can be diagnosed;

d) kidney malignant tumors cannot be distinguished from the rarely inflammative pseudo tumors after the seal of perforation. Sometimes it is difficult to differentiate between a kidney with several large cysts and the complete picture of hydronephrosis.

These and other problems indicate that at the present status of sound imaging technique the sound images of space-occupying lesions have to be interpreted in every case in connection with the findings of other diagnostic method, until a better understanding of acoustic histology is achieved. This leads us to the problem of ultrasonic microscopy.

6.5.1 Acoustic microscopy

To reveal the real difference in the information content of a sound image and an optical image of the same target which may lead to a self contained acoustic histology of biological specimens, it would be desirable to have a device that can magnify the sound image of the object revealing new details.

The scattering of the acoustic waves is dependent on the change in elastic properties, whereas it is the change in the refractive index that determines the scattering of optical waves. This fundamental difference suggests that acoustic radiation, responding to structures within the object, may allow one to resolve details which differ from those recorded with optical waves.

Such a device is the acoustic counterpart of an optical microscope, and was first proposed by Sokolov[7], but its realization was first tried by Suckling[326] in the mid-1950s. Fig. 95 shows the block diagram of his system. It consisted of a water-filled Lucite cylindrical column, having a quartz crystal mounted in a plastic disc that formed a watertight seal at the top end of the column. At the opposite end of the cylinder, sonic lenses were housed. The remainder of the column from this point took the form of a cone that was filled with liquid and extended past the lens housing. The x-cut crystal for the insonification of the object was mounted in the base of a Lucite tray that housed the sample under investigation.

The sound image was formed by the lens onto the receiving crystal that was mounted at the top of the cylinder. This crystal was then scanned mechanically by a fine probe to retrieve the sound image. The signal was then phase-compared with a reference signal from the generating source to determine changes in phase occurring within the sample. It is regrettable that Suckling at that time had no knowledge of Gabor's work, because then he could have become the first to record an acoustic hologram. The phase acoustic microscope, as called by Suckling, worked at 9 MHz and had a magnification of about 4.5. Movement of structures or small organisms (fibrillae or daphnia) could be easily seen.

A few years later Dunn and Fry[327] constructed an absorption acoustic microscope. Their method was based on insonifying the object by planar ultrasonic waves and probing through the transmitted wave by a thermoelectric probe. They have achieved at 12 MHz a 75 micrometer resolution. But these experiments have, to our best knowledge, not been continued. The reason for this may be that a relatively long time, 1 sec, was needed for the thermoelectric probe to establish an appropriate equilibrium at *each* image point.

In the mid-1960s an ultrasonic magnifier was constructed to get a magnified sound image from flaws in metals, and it was also tried for medical purposes[328]. As shown

204

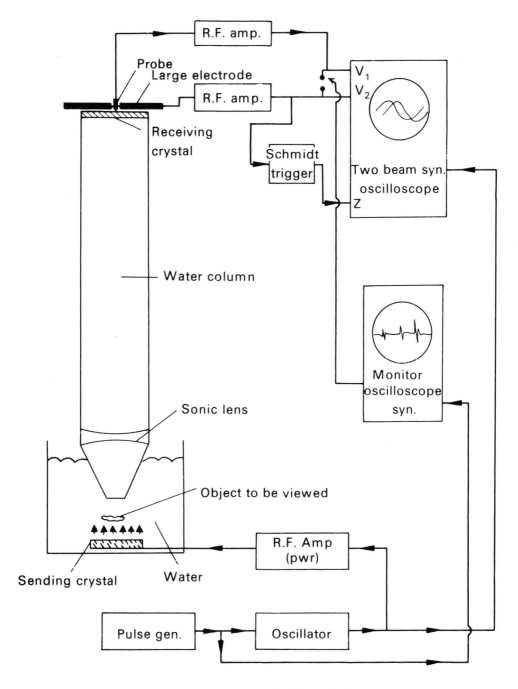

Fig. 95   Block diagram of Suckling's[326] acoustic microscope.

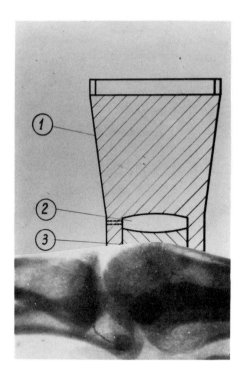

Plate 55 Ultrasonic magnifyer in recording position[328] (Not scaled) 1, aluminum block; 2, mercury lens; 3, face plate.

in Plate 55, it consisted of an aluminium block in which a mercury lens was incorporated. Since the acoustic impedance of mercury and aluminium is nearly the same, there was practically no loss at the boundaries of the acoustic lens, while the refracting index was 4.6. It operated in transmission mode, and the magnified sound image was recorded on sonosensitized or sonosensitive material, but surface levitation technique could also be used to visualize the magnified sound image. Plate 56 shows such a magnified sound image. It represents a hail-shot in a hand. For comparison Plate 57 shows the X-ray image of this hand with the hail-shot. Although the resolution of the magnified sound image is rather poor, it demonstrates the difference between X-ray and ultrasound imaging: in the latter soft tissue is visualized too.

In the 1950s and mid-1960s the possibility of acoustic microscopy was predicted, and with only two problems to be solved:

a) fabrication of ultrasonic transducers for very high frequencies in the order of several hundreds of MHz, and even in the GHz range, which produce a wavelength of 1.5 – 15 micrometer in biological specimens, comparable with biological cell size;

b) purely diffraction–limited acoustic lenses.

In the beginning of the 1970s, the required technological level was reached, and the research to develop a versatile acoustic microscope started in two directions:

a) forming a non sampled sound image and detecting it by one or another method

206

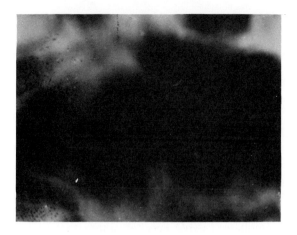

Plate 56   Magnified sound image of a hail-shot in a hand.

discussed in 4.2 and 4.3.1,

b) forming the sound image by mechanical scanning.

Acoustic microscopes belonging to group (a) have the capability of producing both optical and acoustic images simultaneously. While those methods belonging to group (b) may be used both for transmission and reflection mode imaging, and demodulation of the image signal can be performed either by amplitude or phase-sensitive detection. Since the competition between the different methods still holds on, it appears meaningful to illustrate very briefly those techniques which may be interesting in acoustic microscopy today.

6.5.1.1   Acoustic microscopy based on particle orientation   Acoustic microscopy based on *particle orientation* was proposed first by Cunningham and Quate[329].

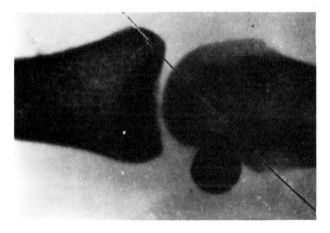

Plate 57   X-ray image of the hand with the hail shot.

207

According to this suggestion the sound image detector is a thin layer of fluid in which 1 micrometer diameter latex spheres are suspended. Although this layer is separated from the target under investigation by a thin membrane, the radiation forces accompanying the sound transmitted through the object will cause local concentration of spheres, which show-up in optical density changes, and the acoustic image may be viewed through an optical microscope. If an acoustic reference beam is employed, an acoustic hologram can be formed, but in this case only a subsequent optical reconstruction will yield a sound image. The resolution obtainable with non holographic technique was about 9 micrometer at 1.1 GHz, but the optical reconstruction of the hologram yielded a resolution of several magnitudes lower.

    6.5.1.2 Acoustic microscopy based on Bragg diffraction Acoustic microscopy based on Bragg diffraction was first proposed by Havlice *et al*[330]. A YAG/mercury acoustic lens was used which Fourier-transformed the sound field, scattered by the object, into a lithium niobate (LiNbO₃) crystal, where it interacted with a laser beam. The Bragg-scattered light was separated from the incident light by polarization, and the acoustic image was optically re-transformed at the image plane. The resolution achieved so far with this technique is in the order of 22 micrometer at about 900 MHz.

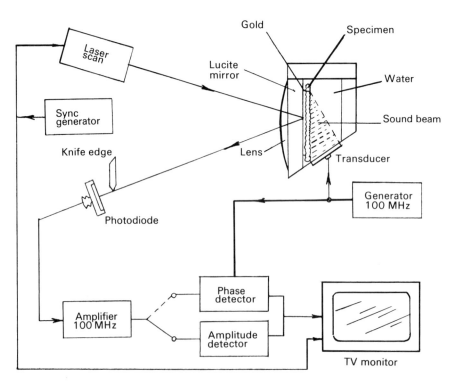

Fig. 96 Acoustic microscope based on solid surface deformation detection, after Korpel *et al*[331].

208

6.5.1.3 Acoustic microscopy based on solid surface deformation detection Solid surface deformation detection was first constructed by Korpel *et al*[331]. As shown in Fig. 96, sound image is projected onto a semireflecting coating of a plastic mirror which becomes spatially modulated by the localized surface deformation produced by the sound image. This is detected by measuring the amplitude of the spatial modulation imparted to the reflection of the scanning laser beam, and the sound image is then represented on a TV screen. Employing an acoustic reference signal an acoustic hologram can also be displayed on the screen.

Since a part of the scanning laser beam is transmitted through both the mirror and the specimen, it can be picked up by a photodiode, and the optical through-trans-mission image of the specimen can be displayed. The specimen has to be very thin, due to the very high sound absorption at those very high frequencies. Resolutions in the order of 10 micrometer have been achieved at about 220 MHz.

To our best knowledge, this is the only acoustic microscope system which is commercially available[332]. This so-called sonomicroscope operates up to 500 MHz. The acoustic image is detected at approximately 40 000 points, and for comparative

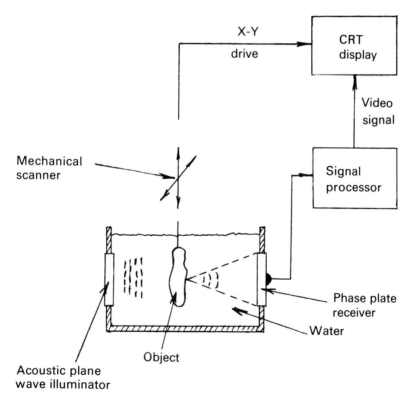

Fig. 97 Schematics of an acoustic microscope after Auld and Farnow[333].

209

purposes the optical image of the sample is simultaneously displayed on a second TV monitor. Changes in acoustic index of refraction resulting from density and elastic variations within the sample can be displayed in an acoustic interferogram mode without the need for repositioning the sample.

6.5.1.4 Acoustic microscopes based on acoustic lens sampling Acoustic microscopes based on acoustic lens sampling use either electronically or by refraction focusing lenses to build up the scanned image. In their setting up they are close relatives to the scanning electron microscope, only not the interrogating beam, but the target under investigation is moved.

An electronically focused Fresnel-zone plate transducer as described in 4.3.5.1 has been successfully used for acoustic microscopy by Auld and Farnow[333]. As shown in the schematics of Fig. 97, the trans-insonified target under investigation is mechanically raster scanned in a water bath in the focal plane of the electronically focused transducer. The mechanical motion is synchronized with a raster on the display, and since the transmitted power modulates the intensity of the electron beam of a CRT, the sound image representing the acoustic transmissibility of the target will be visualized on the screen. The resolution in water at about 10 MHz is in the order of less than two wavelengths. The scanning acoustic microscope seems to become a hard competitor to the acoustic microscope described in 6.5.1.3, although to our best knowledge, it is not yet commercially available.

As seen from the schematics of Fig. 98 the basic idea of the scanning acoustic microscope, which was introduced by Lemons and Quate[334] is rather simple. The

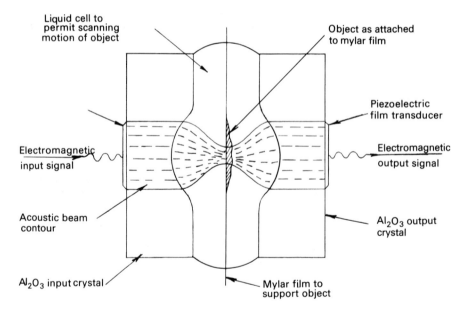

Fig. 98 Schematics of a scanning acoustic microscope after Lemons and Quate[334].

210

specimen is placed on a suitable target holder in the joint focal plane of two acoustic lenses. To eliminate standing waves, the insonifying beam, which is focused on the target by one of the lenses, is a gated-sine-wave with a frequency of up to 1000 MHz. Since the attenuation of sound increases considerably as the frequency is increased, a compromise between good resolution and easy operation has to be made. Attenuation, for instance, in water at 1 GHz is about 200 dB/mm. It was found that for most practical applications the resolution offered by 300 MHz is adequate, and in this case the attenuation will be only 18 dB/mm.

The transmitted energy which diverges from the acoustic focus is collected and collimated in the receiver crystal by the second cofocal lens. This symmetry of the lenses has the advantage of enhancing the resolution of the system by a factor of about 2, and also of improving the signal-to-noise ratio. Since there is no background signal, the information that is detected is limited to that coming from the focus. This, however, indicates that the most essential part of this microscope is the lens-transducer system. Since single surface lenses are used, based on our optical experience, we are inclined to think that the spherical aberrations inherent in such single surface lenses would limit the waist of beam to a diameter of several wavelengths, and the resolution power would be seriously degraded. Spherical aberration is, however, inversely related to the ratio of the wave velocities, the typical value of which is in optics of about 1.5. In ultrasonics it may be over 7, which means that spherical

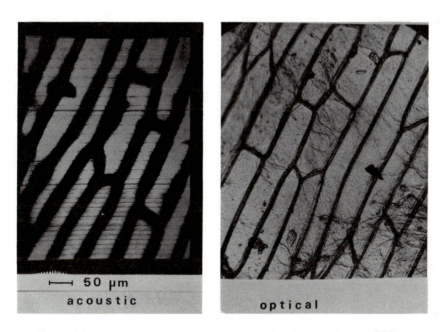

Plate 58 Micrographs of an onion skin recorded by acoustic and optical microscopes[335].

211

aberrations in acoustic lenses are almost two orders of magnitude smaller than in optical ones.

Lemons and Quate used acoustic lenses made from sapphire crystal, while Penttinen[335] has used recently a YAG (yttrium-aluminum-garnet) rod, which is easier to handle, although its wavelength ratio to water is somewhat less than that of sapphire crystals (5.73 and 7.46, respectively).

The pixel information is picked up by the other ultrasonic lens and is converted into an electric signal which modulates the image point on the CRT screen. Since the movement of the target is performed in a raster pattern, which is synchronized with the sweep at the CRT, the transmission sound image of the target will be visualized on the display. Plate 58 shows the familiar onion skin sample, where the optical and acoustic images are inverted with respect to each other along the diagonal of the picture, recorded by the scanning acoustic microscope described in [334] and [335].

The scanning acoustic microscope can, however, operate in reflection mode too, but then only a single lens is used. The receiver unit must be connected to the transmitting transducer lens system, which can be accomplished, for instance, by a directional coupler or a hybrid junction that isolates the transmitted pulse from the received one. Plate 59 displays an acoustic and an optical image of a transistor planar structure where the acoustic image was recorded in a reflection mode. Since in this case the image information was led to the y axis of the display, and the y scan of the object was directed along the diagonal axis in the display, a similar isometric 3-D like image was obtained, as is usual in scanning electron microscopy. The resolution in this case is 50 micrometre per division.

The technique of scanning acoustic microscopy can further be extended to techniques analogous to dark-field and stereo-microscopy in optics using a cofocal arrangement where the angle between the axis of each lens can be continuously varied. In this arrangement the effective depth of focus is diminished, and the higher spatial frequencies of the scattered radiation are detected[336].

There are several unexploited ways in which performance of the scanned acoustic microscope can be improved. So if the output circuit to the second harmonic of the input signal is tuned and the scanned image is recorded at this double frequency, the information will be different from that obtained in linear imaging process. It provides information on the nonlinear response of the target investigated to acoustic waves. This may be very important in biological research[337].

6.6   And psychophysics

In the course of discussing methods and techniques how to process sound-borne information patterns in a form we are accustomed to when light-borne information is processed, that is in a two-dimensional pattern called image, we have emphasized several times that even if the sound image is visualized and we can perceive it with our vision, it will have a meaning only if our eye-brain system has an *a priori*

acoustic ┣━━┥ 50 µm

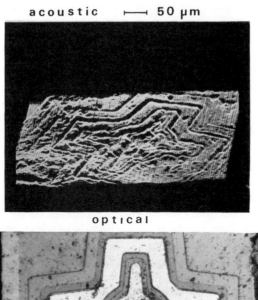

optical

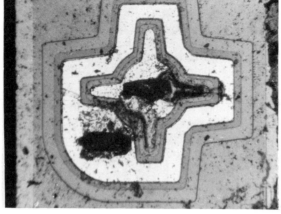

Plate 59 Micrographs of a transistor planar structure recorded by acoustic and optical microscopes in reflection mode[335].

knowledge to evaluate it. In this evaluation the laws of psychophysics play an important role. It has, however, never been questioned whether it is really needed that the information pattern conveyed by sound, i.e. by a mechanical wave, be converted first into an information pattern which is conveyed by electromagnetic waves to perceive a sound image.

Why question it? If acoustical-to-optical conversion were not needed, even blind people could solve perhaps the problem of locating and identifying desired objects or they would get pictorial information as to shapes and details of objects in the extracorporal world via sound waves.

The answer to this question is not a simple one if we consider a little the evolution of biological information processing, since it states that the ability of an organism to

213

process information patterns conveyed by mechanical waves preceded the ability to process information patterns conveyed by electromagnetic waves, that is palpitation and hearing preceded vision. Although each of these information processing abilities followed a different direction of evolution in different species, information borne by mechanical waves was used for short distance information exchange, and vision for long distance.

Perceiving an information pattern, however, is more than witnessing a signal pattern. According to MacKay[339] perception is an internal adaptive reaction by way of receptor organs to the demands made by the world, or an internal 'updating' of the organizing system to match incoming signals. Referring this to electromagnetic information pattern recognition we can say that the optical image on the retina as a receptor field is transformed from a continuum of light and distributed onto a mosaic of luminous points from which the visual world to be acted upon is extracted[340]. In order to do this it is necessary to decide which points go together, which form a unit object, which constitute the boundaries that delineate it, and which constitute the background. This can be done in many ways, depending on the aspects of information pattern that are significant. Lower animals have more crude and rigid processing schemes than man does. The neural activity of man has a far higher plasticity and so he has the ability of learning the 'visual' perception of 'non adequate' stimuli.

What now has been said on the perception of an optical image has to be valid also for the perception of sound images, only during the evolution of mankind this ability of the organism was put into the shade. In the animal world, however, it still exists, since there are different species which can distinguish between objects of different shapes, based on the information pattern conveyed to them by mechanical waves[1, 2].

The overshadowing of the ability of sound image perception during the evolution of mankind can well be understood if we consider that 'image resolution' offered by the wavelengths belonging to the audible frequency range could not offer an advantage over that of an optical image. (The changing of the perceivable frequency range to shorter wavelengths during evaluation and to get by this a better sound image would not have been a solution, since due to the rapidly growing sound absorption with frequency[3.4], the range of its applicability would have been considerably decreased.) The capability of processing sound images, however, has most probably not disappeared completely, and if a sound image were presented in an adequate form, man could perhaps learn to perceive it as if it were an optical image. The emphasis is on the 'adequate' presentation of the sound information pattern, and both the ear and the skin have to be considered as a counterpart of the retina. This means further that resolution of the presented sound image has to be comparable to the crude optical image and be still enjoyable. Only sound patterns formed by ultrasonic waves can be really considered. This, however, involves the problem of transforming ultrasonic image into a form in which it is perceivable to the ear or the skin.

214

6.6.1 Seeing by sound as delphinids do

Delphinids or other echolocating animals can discriminate between targets of different shapes such as circles, squares, triangles, etc., just by evaluating the sound pattern reflected from the target. The recent experiments of Fish *et al*[341] demonstrated that human divers can perform sonar discrimination with a skill similar to that of delphinids, documenting that the ability of the human information processing mechanism to process two-dimensional sound information pattern has not disappeared completely.

In these experiments the human divers were wearing helmets with one sending and two receiving transducers, as shown in Plate 60. Porpoise-like broadband clicks, with energy centered about 60 kHz, were projected and the returned echoes were digitized, and after filtering and amplifying they were presented at a rate 128 times slower than the record rate. The minimum time duration required for humans to discriminate between complex transients is about 2 msec, and so 20 μsec short echoes could be processed by the diver. In order to play the stretch echoes back to the diver before the next pulse was emitted, the pulse repetition rate had to be set to 15.6 Hz.

Pairs of targets, previously used in delphinid sonar experiments were presented to divers, and one of the two targets was always the same 'standard' target. There was

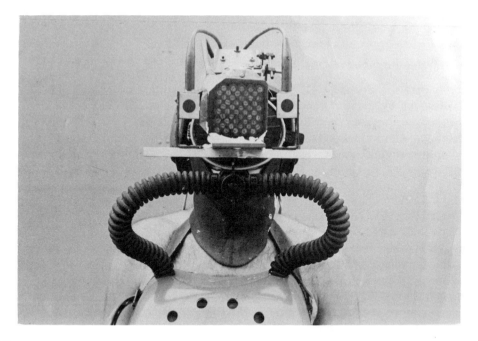

Plate 60 Diver using a helmet with one sending and two receiving transducers, 11.5 cm apart (*courtesy* of Official Photography US Navy).

215

no restriction placed on the diver's movements, and it was observed that, particularly during difficult discriminations, the divers were swinging their heads similar to that described by Evans and Powell[342] in their porpoise experiments, so that, in both cases the target space is scanned. It is, however, known that in the evaluation of a two-dimensional optical information pattern self-generated motion is always involved[343], thus this finding underlines the similarity between sound image and optical image perception. That in sound image perception, not scanning as such, but *self-generated* scanning, i.e., the immediate feedback of the sound of the echo as a function of the head position, is important is backed up by the fact that a person outside the tank receiving the same sound information pattern and watching the diver had difficulty in determining how discriminations were made even on targets that produced very different sound patterns.

Fish *et al*[341] report only on target discrimination, and it is not known whether the divers during discrimination had also an 'image-like' impression from the target, but if not, this means only that they have to learn to 'see' the perceived two-dimensional sound information pattern. This is, however, not at all strange, since learning is a critical factor also in the ability of seeing an optical image, as Gibson[344] has pointed out: 'the *invariants* in a pictorial array, the information *about* the dog, cat, man, house or car, are picked up very early, but the embodiment of the . . . shape of an object is not noticed, and children do not take a pictorial attitude toward ordinary objects until they learn to do so.'

6.6.2   *Seeing by sound via the skin*

The pioneer work of Bach-y-Rita[345] on sensory substitution may open quite a new field in seeing by sound. He presented evidence that central perceptual mechanism can adjust to a sensory input originating in an artificial receptor. Delivering to the skin the spatial information pattern gathered by a television camera through an array of vibrating stimulators or electrodes, blind persons could identify and correctly locate in space forms, objects, figures and faces, after a shorter or longer training. It has to be emphasized that the subjective localization of the light wave conveyed spatial information was not on the skin but in the three-dimensional space, independently of the place where the skin stimulator matrix was put. Already a matrix of 20 × 20, a 400 point system, provides enough information even to the congenitally blind, so that he could see and discriminate between different objects, and he could describe the shape of them, etc. The reason for this is that there is no need for complex topological transformation or temporal coding for the direct presentation of pictorial information onto the accessible areas of the skin.

In 4.3.4.1 we indicated that with a 16 × 16 matrix of 2N electret foil array, recognizable and identifiable sound images can be obtained. This, however, means that if such a sound imaging detector array and a skin surface stimulator array were combined, an investigator could 'see' a pictorial impression of the insonified target

216

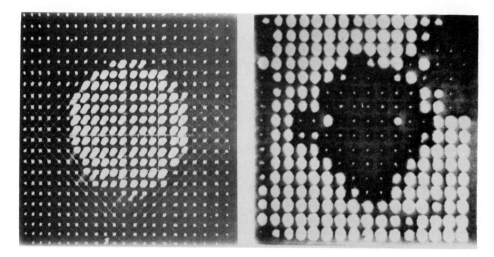

Plate 61 Left side: stimulus pattern of a disc appearing on a tactile array[345], right side: acoustic shadow of a metal plate appearing on an electret foil array. Dark area in the top right corner is due to nonuniformities of the soundfield[140].

in the same fashion as if the target was illuminated and looked at with the eyes.

The left side image of Plate 61 shows the stimulus pattern of a disc appearing on the tactile array, while the right side image is in the acoustic shadow of a metal plate appearing on an electret foil array. In both cases the highly illuminated arrays are activated area elements.

Naturally, sound image seen this way will have a lower resolution than the optical one would have, since the information bearing wave is larger. We, however, do not know yet whether this wavelength-introduced 'roughness' of the perceived sound image would be made conscious or not, since, as Eccles[346] pointed out, patterned visual perception is not inborn: it is an interpretation of the retinal data that has been learned to accomplish in life. Consequently, not only the language of the optical world, but the language of the sonic world also has to be learned. This, when accomplished, will surely augment positively the capacity of man to understand the world in which he lives[347–351].

References

[248]TUCKER, D. G. (1965) *Sonar in Fishery* Fishing New Books Ltd. London

[249]KONDOH, (1977) *J.E.E.* = April 43–45

[250]GREISCHAR, L. L., CLAY, C. S. (1972) *J. Acoust. Soc. Am.* **51** 1073–1075

[251]HARGER, R. O. (1970) *Synthetic Aperture Radar System* Acadmic Press, New York

[252]U. S. Pat. 3,914,730 (1975) Ch. H. Jomes, N. Wolfe

[253]KOCK, W. E. (1973) *Proc. IEEE* **61** 1518–1519

[254]CUTRONA, L. J. (1975) *J. Acoust. Soc. Am.* **58** 336–348

[255]WHIPP, E., HORNE, D. A. (1976) *Ultrasonics* **14** 201–204

[256]WILLIAMS, R. E. (1976) *J. Acoust. Soc. Am.* **60** 60–73

[257]JONES, R. A., PRESTON, K., JR. (1975) *IEEE Trans on Circuits & Systems* **CAS-22** 293–303

[258]CUTRONA, L. J. (1970) *Synthetic Aperture Radar. Radar Handbook* ed. M. Skolnik, Chapter 23 McGraw-Hill, New York

[259]JONES, C. H., GILMOUR, G. A. (1976) *J. Acoust. Soc. Am.* **59** 74–85

[260]ROLLE, A. L., WERLE, C.R.T. (1971) *Naval Coastal Systems Lab. Report* NSRDL/PCC 3485

[261]U.S. Pat. 3,506,952 (1967) D. Gabor, B. B. Bauer, F. B. Gorman

[262]FARRAH, E., MAROM, E., MUELLER, R. K. (1970) *Acoustical Holography* Vol. 2 ed. A. F. Metherell, pp. 173–183 Plenum Press, New York

[263]KAMATA, H., OHGAKI, M. (1977) *J. Acoust. Soc. Japan* **33** 190–197

[264]FARHAT, N. H. (1972) *Proc. Symp. Engineering Applications of* Holography 295–314

[265]BOUTIN, H., MAROM, E., MUELLER, R. K. (1967) 74th Meeting of the Acoustical Society of America

[266]URICK, R. J. (1967) *Principles of Underwater Sound for Engineering* McGraw-Hill, New York

[267]SHIBATA, S., KODA, T., NATSUMOTO, S., YAMAGA, J. (1977) *J. Acoust. Soc. Am.* **62** 819–824

[268]SHIBATA, S., KODA, T., YAMAGA, J. (1978) *Ultrasonics* **16** 65–68

[269]CLAERBOUT, J. F., SCHULTZ, P. S. (1975) *Proc. Image Processing for 2-D and 3-D reconstruction from Projections* Stanford

[270]U.S. Pat. 3,400,363 (1968) and 3,450,225 (1969) D. Silverman

[271]U.S. Pat. 3,729,704 (1973) J. B. Farr

[272]U.S. Pat. 3,852,709 (1974) R. K. Mueller

[273]SATO, T., SASAKI, K. (1977) *J. Acoust. Soc. Am.* **62** 404–408

[274]HAYSER, R. C. (1967) *J. Acoust. Eng. Soc.* **15** 370–382

[275]U.S. Pat. 3,466,652 (1969) R. C. Hayser

[276]LECROESETTE, D. H., HAYSER, R. C. (1972) *3rd Conf. Med. Physics* Göteborg, Sweden

[277]JACOBS, J. E., MALINKA, A. V., HAQUE, P., IHABUALA, M. (1976) *Ultrasonics* **14** 84–90

[278]STROKE, G. W., FUNKHOUSER, A. T. (1965) *Phys. Lett.* **16** 272–274

[279]GREGUSS, P., WAIDELICH, W. (1975) *IEEE Trans. on Computer* **C-24** 412–418

[280]SENAPATI, N., LELE, P. P., WOODIN, A. (1972) *Proc. IEEE Ultrasonic Symp.* 59–63

[281]PIMMEL, R. L. (1972) *Ultrasonics* **10** 262–266

[282]NICHOLAS, D., HILL, C. R. (1975) *Proc. Symp. Ultrasonics Internat'l.*

[283]CHIVERS, R. C., HILL, C. R. (1975) *Phys. Med. Biol.* **20** 799–815

[284]HILL, C. R., NICHOLAS, D., BAMBER, J. C. (1976) *7th L. H. Gray Conf. Medical Images*

[285]ALDRIDGE, E. E., SAWYERS, J. F. (1966) *Ultrasonics* **4** 131–135

[286]D. Pat. 2,605,776 (1976) P. Greguss

[287]KRAUTKRÄMER, J., KRAUTKRÄMER, H. (1969) *Werkstoffprüfung mit Ultraschall* Springer Verlag, Berlin Göttingen, Heidelberg

[288]SPROULE, T. (1969) *Ultrasonics for Industry Conf.* 56–59 IPC Science and Technology Press, Guildford

[289]HAROLD, S. O. (1972) *A Solid State US Image Converter* Report of Electrical and Electronic Engineering Research Reports Porthmouth Polytechnique

[290]LANDRY, J., WADE, G. (1972) *SPIE Proc.* **29** 47–54

[291]COLLINS, H. D., GRIBBLE, R. P. (1972) *SPIE Proc.* **29** 67–82

[292]PAPADAKIS, E. P. (1974) *J. Acoust. Soc. Am.* **55** 783–784

[293]CURTIS, G., JOINSON, A. B. (1975) *Proc. Ultrasonics International* 156–161, IPC Science and Technology Press, Guildford

[294]CURTIS, G. (1975) *Ultrasonics* **12** 148–154

[295]SPEAKE, J. H., CURTIS, G. (1977) *Proc. 9th ICA* K60

[296]DUSSIK, K. T. (1942) *Z. Ges. Neurol. Psych.* **174** 153–168

[297]LUDWIG, G. D., STRUTHER, F. W. (1947) *Naval Medical Research Inst. Project* NM004:001 4 1–23

[298]WHITE, D. N. (1970) *Ultrasonic Encephalography* Hanson & Edgar, Kingston

[299]*Proc. Ultrasonic Medical Diagnostic Symposium* ed. L. Adler, The University of Tennessee, Knoxville (1975)

[300]WHITE, D. N., BROWN, R. E. (1977) *Ultrasound in Medicine* Plenum Press, New York

[301]MAKOW, D. M., WYSLOUZIL, W., WHITE, D. N., BLANCHARD, J. (1966) *Acta Radiologica* **5** 855–864

[302]WHITE, D. N., CLARK, J. M., CHESEBROUGH, J. N., WHITE, M. N., CAMPBELL, J. K. (1968) *J. Acoust. Soc. Am.* **44** 1339–1345

[303]SATO, T., WADAKA, S. (1975) *J. Acoust. Soc. Am.* **58** 1013–1017

[304]SATO, T., WADAKA, S., ISHI, J. (1977) *J. Acoust. Soc. Am.* **62** 102–107

[305]WHITE, D. N. (1972) *Compt. Biol. Med.* **2** 273–284

[306]Diagnostics Electronics Co., Lexington, Massachusetts

[307]BUSCHMANN, W. (1966) *Einführung in die ophthalmologische Diagnostik* G. Thieme, Leipzig

[308]GITTER, K. A., KEENEY, A. H., SARIN, L. K., MEYER, D. (1969) *Ophthalmic Ultrasound* The Mosby Co., St. Louis

[309]SUSAL, A. L. (1974) *Ultrasonics* **12** 36–39

[310]GREGUSS, P., BERTENYI, A. (1968) *Acta Fac. Med. Univ. Brunensis* **35** 133–136

[311]CHIVERS, R. C. (1974) *Ultrasonics* **12** 209–213

[312]ALDRIDGE, E. E., CLARE, A. B., SHEPHERD,

D.A. (1974) *Ultrasonics* **12** 155–160

[313]TRIER, H. G. (1977) *Gewebedifferenzierung mit Ultraschall Bibliotheca Ophth.* No. **86** S. Karger, Basel.

[314]ASBERG, A. (1967) *Ultrasonics* **5** 113–117

[315]FEIGENBAUM, H. (1972) *Echocardiography* Lea & Febiger, Philadelphia

[316]WAAG, R. C., GRAMIAK, R. (1974) *Proc. IEEE Ultrasonic Symp.* **896–1 SU** 12–15

[317]YUSTE, P. (1975) *Ecocardiografia* Interamericana SA, Madrid

[318]HOUSTON, A. B., GREGORY, N. L., COLEMAN, E. N. (1977) *Brit. Heart J.* **39** Sep.

[319]Searte Ultrasound, Santa Clara, California.

[320]BROWN, T. G., YOUNGER, G. W., SKRGATIC, D., FORTUNE, J. (1977) *Ultrasound in Medicine* ed. D. N. White, R. E. Brown pp. 1797–1799 Plenum Press, New York

[321]GREGUSS, P. (1977) *J. Computer Assisted Tomography (Computed Tomography)* **1(2)** 184–186

[322]WATERS, J. P. (1972) *Appl. Phys. Lett.* **9** 405–407

[323]WILD, J. J., REID, J. H. (19) *Am. J. Path.* **28** 831–861

[324]KOBAYASHI, T. (1977) *Radiology* **122** 207–214

[325]LUTZ, H., PETZOLDT, R. (1976) *Ultrasonics* **14** 156–160

[326]SUCKLING, E. E. (1956) Ph. D. Thesis, Polytechn. Institute of Brooklyn, Publ. No. 18356

[327]DUNN, F., FRY, W. J. (1959) *J. Acoust. Soc. Am.* **31** 632–634

[328]GREGUSS, P. (1965) *Wiss. Z. Humboldt-Univ. Berlin, Math.-Nat. R.* 14 155–156

[329]CUNNINGHAM, J. A., QUATE, C. F. (1972) *Acoustical Holography* Vol. 4 ed. G. Wade pp. 51–71 Plenum Press, New York

[330]HAVLICE, J. E., KAMPFER, R., QUATE, C. R. (1970) *Progress Towards an Acoustic Microscope* Stanford University ML Report 1928

[331]KORPEL, A., KESSLER, L. W., PALERMO, P. R. (1971) *Nature* **232** 110–111

[332]Microsonoscope, Sonoscan, Inc. Bensenville, Illinois.

[333]AULD, B. A., FARNOW, S. W. W. (1974) Hansen Laboratory of Physics, Stanford University Annual Report **ML** 2358

[334]LEMONS, R. A., QUATE, C. F. (1974) *Appl. Phys. Lett.* **24** 163–165

[335]PENTTINEN, A., LUUKKALA, M. (1977) *Ultrasonics* **15** 205–210

[336]PENTTINEN, A. (1977) *Ph. D. Thesis* Department of Technical Physics, Technical University, Helsinki

[337]BOND, W. L., CUTLER, C. C., LEMONS, R. A., QUATE, C. F. (1975) Appl. Phys. **27** 270–272

[338]KAMPFER, R., LEMONS, R. A. (1976) *Appl. Phys. Lett.* **28** No. 6

[339]McKAY, D. M. (1965) *Prog. Brain Research* **7** 321–332

[340]MATURANA, H. R. (1964) *Int. Congr. Ser.* No. **49** pp 170–178 Excerpta Medica Foundation, Amsterdam

[341]FISH, F. J., JOHNSON, C. S., LJUNGBLAD, D. K. (1976) *J. Acoust. Soc. Am.* 59 602–606

[342]EVANS, W. E., POWELL, B. A. (1967) *Animal Sonar Systems* ed. R. G. Busnel, Laboratoire de Physiologie, Jouy-en-Josas

[343]JARBUS, A. L. (1967) *Eye Movements and Vision* Plenum Press, New York

[344]GIBSON, J. J. (1966) *The Senses Considered as Perceptual Systems* Houghton, Boston

[345]BACH-Y-RITA, P. (1972) *Brain Mechanisms in Sensory Substitution* Academic Press, New York

[346]ECCLES, J. C. (1965) *The Brain and the Unity of Conscious Experience* Cambridge University Press, London New York

[347]GREGUSS, P. (1979) Proc. 2nd Meeting of WFUMB. *Excerpta Medica*, Amsterdam (In Press)

[348]GREGUSS, P. (1979) *Ultrasonics International '79*, Graz

[349]GREGUSS, P. (1979) *Proc. Animal Sonar Systems Symposium* ed. R. G. Busnel, Plenum Press, New York (In Press)

[350]GREGUSS, P. (1979) *Meeting of the Hungarian Biophysical Society*, Tihany

[351]GREGUSS, P. (1979) *UBIOMED IV* Visegrad

Glossary

Acoustic image   two-dimensional amplitude distribution of a sound field of any frequency, but not necessarily visualized.

AHFS   Acoustic Holographic Fourier Spectrometer.

alpha numeric   capable of using both letters and numbers.

AOCC   Acoustical-to-optical Conversion Cell using nematic liquid crystals.

area sensor   device yielding directly two-dimensional information pattern.

central slice theorem   a mathematical method allowing the formation of a cross-section image from projections of three-dimensional objects.

CFRP   Carbon Fibre Reinforced Plastic.

echocardiography   ultrasonic heart diagnosis with A- and/or M-mode display.

echoencephalography   ultrasonic brain diagnosis with A-mode display.

Fresnel-zone pattern   a ring pattern showing decreasing spacing with increasing distance from the centre. The nth zone is located at a radius r_n given by $r_n = r_1 \sqrt{n}$, where n = 1, 2, . . .

hyperphonogram   Term introduced by Dussik brothers for transmission sonograms of the head.

information capacity   measure of optimum coding.

information content   result achieved under practical conditions.

insonification   acoustic equivalent of optical illumination.

isometric image   picture characterized by equality of measures.

microencapsulation   embedding liquid, solid or gaseous particles in compartments varying from a few micrometres to about 1000 micrometres.

neoplastic   constituting of a new growth of tissue serving no physiological function, e.g. a tumor.

Nyquist criterion   to detect a spatial frequency it is necessary to have at least two sampled points within each cycle.

optical replica of a sound image   optical point-by-point correlation of the two-dimensional intensity distribution of a sound field.

orthoscopic   giving an image in correct and normal proportions.

pixel   image element for which the point spread function is reduced to 10% of its peak value.

projection slice theorem   central slice theorem.

RUSA   Recording Ultrasonic Spectrum Analyzer.

Schlieren method   a visualization method using light streaks resulting from the change of refractive index of the medium.

SEA   Sound Emission Analysis.

sonic image   acoustic image.

sonogram   generally A-mode display, but sometimes used in the sense of acoustic image.

sonosensitive plate   counterpart of photographic plate, sensitive to sound waves.

speckle pattern   random intensity distribution resulting from fairly coherent optical illumination or insonification.

TFT-EL display   Thin Film Transistor Electro Luminescent solid-state display.
tomography   recording only one 'slice' of object space.
typography   the art of letter pressing.
UTTAR   Ultrasound Transmission Tomography of Attenuation by Reconstruction.
UTTVR   Ultrasonic Transmission Tomography of Velocity Reconstruction.

INDEX

Huygen's principle 30

image quality 16
impedance of plane waves 29
information theory 16–25
information and sound characteristics 36–40
interlace scanning method 97
interferometer, Twyman Green type 89
isochronicity 143–5
isometric imaging 149–50

Langevin, P. 12
Lamb waves *see* plate waves
laser beam sampling 76–9
lens, electronic focusing 104–8
levitation of liquid surfaces 61–3, 125–7
linearity 143–5
linear scanning 80–82, 95
liquid crystals 14, 53–72

mechanical sampling 79–85
medical applications 189–212
   and space-occupying lesions 202–4
microencapsulation 46
microscopy 204–212
modulation transfer function 75–6
multiplexing, holographic 133–6

nematic liquid crystals 56–9
Newtons rings 47
non-destructive testing 182–9
non-sampled sound images 42–54
Nyquist criterion 74–5

Ohm's law of acoustics 29
oscillatory transducer 81–4

particle displacement 26–7
passive sound imaging 187–9
penetration at boundaries 33–4
perception of sonic images 37–8
phase 26
photoelasticity 55–6
photographic plates, sonosensitized 42–7
   and image formation 44
piezoelectric effect 12
piezoelectric area detector 79–80, 86, 126
piezoelectric array 91–2
plate waves 28
Pohlman cell 60
point-by-point insonification 73
polyscan 184–7

pressure amplitude 29
propagation, laws of 30
psychology and perception 212–17

quantization 73

Raman Nath cell 64
Rayleigh resolution criterion 16
raster point 73
raster line 73
recognition of target 113
recording levels 22–3
recording ultrasonic spectrum analyser 179–82
reflection, coefficient evaluation 32–4
   and spatial information 39–40
reflection imaging and Bragg diffraction 67
resolution 16, 101–2
rotating scanners, underwater 161–2
rotating transducer 81

sampling 72–6
   electron beam 86–90
   electronic 90–8
   laser beam 76–9
   linear 80–2
   mechanical 79–85
   sector 82–5
      with acoustic lens 98–108
scanned deflection modulated image 117
scanned intensity modulated image 119
schlieren technique 14, 54–5
shadow casting 39
Shannon's theorem 19–20
signal-to-noise ratio 18–19
Snell's law 34
Sokolov, S. Y. 12
Sokolov's liquid surface levitation method 61
solid surface deformation 63–5
   and laser beam scanning 76–9
sonochemical reactions 45–6
sonochromes 14, 45–6
sonoelectrochemical reactions 47–8
sonoluminescence 48–52
sonosensitized plates 42–4, 127
   theoretical resolution 43
sound characteristics 26–7
   and information 36
sound field 29–30
sound images, non-sampled 42–54
   sampled 72–6

223